THE GREAT EVOLUTION MYSTERY

THE GREAT EVOLUTION MYSTERY

Gordon Rattray Taylor

A Cornelia & Michael Bessie Book

HARPER & ROW, PUBLISHERS, New York
Cambridge, Philadelphia, San Francisco, London
Mexico City, São Paulo, Sydney

 For information address Harper & Row, Publishers, Inc., 10 East 53rd Street, New York, N.Y. 10022.

ISBN: 0-06-039013-1

LIBRARY OF CONGRESS CATALOG CARD NUMBER: 82-47535

Contents

Foreword

For over two years Gordon worked on this book, battling all the time against an illness which grew steadily more serious and debilitating. It was a heroic effort, and, when he died in December 1981, the book was complete but unrevised.

In this work he has attempted a synthesis of the arguments now raging among the students of Evolution, and has explained recent work on genetics and on the origin of 'non-equilibrium systems' which throw a new light on this century-old subject.

I am deeply grateful for the invaluable help and encouragement I have received in trying to carry out revisions, to find illustrations, to compile a bibliography, and so on. I would like to thank particularly Professor Avro Mitchison of University College, London, Dr Colin Patterson of the British Museum (Natural History) and Dr Bob Savage of Bristol University who read the manuscript. They are of course not responsible for errors which remain.

I would also like to thank John Dawes of Bath University who helped with the glossary, Sue Moore whose help as editor was so important, Caroline Douglas-Cooper and my stepdaughter, Michele Lewars, for whose help in research and typing I am deeply grateful, and, above all, Tom Rosenthal of Secker and Warburg, most generous of publishers, who kept faith with Gordon and supported him staunchly through the difficult years of his illness.

Olga Rattray Taylor
Bath
1982

Introduction

Evolution in Ferment

Darwin's theory of evolution by natural selection, which has stood as the one great biological law comparable with the laws of physics for more than a century, is crumbling under attack. Biologists are discovering more and more features which it does not seem able to explain, and are holding meetings at which tempers often run high to discuss problems, some new, others which were discovered decades ago but quietly ignored.

If Darwinism fails, this is a critical point in human rationality, of much more than merely scientific consequence. The reason Darwin's ideas caused such a furore when they were first announced was that they presented the living world as a world of chance, determined by material forces, in place of a world determined by a divine plan. They substituted chance for purpose. They removed evolution from the metaphysical to the material.

While for some people it is more comforting to believe that everything is in the hands of a deity, a paternal figure who will see all is well provided we submit to him, others find this stultifying and very much prefer to think that we live in a world which we, by our own efforts, can alter. Thus the issue underlying any debate about Chance vs Purpose is, or can be seen as, a political one: Left vs Right. At the same time, there is something repellant about the idea that we live in a meaningless, fortuitous world. Darwinism, therefore, has always aroused fierce feelings and has been attacked and defended with great ferocity.

My purpose therefore is to examine Darwinism and its modern form, neo-Darwinism, in order to see where we stand. To say that Darwin's theory is inadequate or incomplete is not to say that we have to retreat to a mystical explanation of the living world. I believe there is a third alternative. It has constantly happened in the history of science that issues have been presented as straight choices but have turned out to be neither one nor the other, but an amalgam of both. The transmission of signals in the nerves, for instance, was seen as electrical by some neuroscientists, as chemical by others. It turned out to be a bit of both, and in some ways neither, for both the chemical and electrical aspects were unusual.

I shall suggest that the world is ordered, but that the order springs from an inner necessity. Only a restricted number of paths of future development are open – that is, chance is restricted – and as time passes, the number of possibilities becomes fewer. Like a steel ball rolling down a pinball machine, the nearer we get to our destination, the more limited the possible outcome becomes.

On one point we must be clear before we start. The fact that an evolutionary process occurred is not in doubt. It is only the mechanism which brought it about which is being questioned. Any rational man who examines even a part of the evidence must be convinced that life-forms started as very simple cells, which associated in more and more complex forms, which branched into many variants, over an extremely long period of time. I have to say this in view of the attacks of those known as Creationists, some of whom deny this on Biblical grounds. Creationists are of several kinds, and not all of them insist on the Biblical account of creation in six days; most, however, claim that the various forms of life appeared thanks to a fiat of the Creator and not by gradual evolution from earlier forms.

Darwin proposed to explain these changes by two main propositions: first, that variations occurred by chance, second, that the more advantageous form would be favoured; competition would eliminate the less efficient. Variation and selection were the two keys. Now, there can be no doubt that selection occurs. Many thousands of experiments and observations have shown it at work. Nevertheless, there are quite a large number of phenomena, mostly structural variations or alternatives, which natural selection seems unable to account for. Is there some other principle at work alongside natural selection? If so, what? Piling up cases where selection *does* work will never prove that it is the unique agency in evolution, though many eminent biologists act as if they thought so.

This is one of the mysteries I shall explore; but much more serious for Darwinism and more central to the philosophical question and our desire to know what kind of world we live in is the proposition that variation depends entirely on chance. While it is not difficult to believe that some small structural change, such as a change in the shape of a bird's bill, occurs by chance, it is very difficult to believe that a complex structure like the eye, which involves many coordinated changes, came about by chance, and especially as it did so

God's creation of the world in six days, as described in the Book of Genesis. From a sixteenth-century Bible printed in France

several times. Darwin himself was flummoxed by this. 'When I think of the eye, I shudder,' he said.

The ascription of all changes in form to chance has long caused raised eyebrows. Let us not dally with the doubts of nineteenth-century critics, however, for the issue subsided. But it raised its ugly head again in a fairly dramatic form in 1967, when a handful of mathematicians and biologists were chatting over a picnic lunch organised by the physicist Victor Weisskopf, who is a professor at Massachusetts Institute of Technology (MIT) and one of the original Los Alamos atomic bomb group, at his house in Geneva. 'A rather weird discussion' took place. The subject was evolution by natural selection. The mathematicians were stunned by the optimism of the evolutionists about what could be achieved by chance. So wide was the rift that they decided to organise a conference, which was called Mathematical Challenges to the Neo-Darwinian Theory of Evolution. The conference was chaired by Sir Peter Medawar, whose work on graft rejection won him a Nobel prize and who, at the time, was director of the Medical Research Council's laboratories in North London. Not, you will understand, the kind of man to speak wildly or without careful thought. In opening the meeting, he said: 'The immediate cause of this conference is a pretty widespread sense of dissatisfaction about what has come to be thought of as the accepted evolutionary theory in the English-speaking world, the so-called neo-Darwinian theory. This dissatisfaction has been expressed from several quarters and is not only scientific.'

One such quarter is the philosophers, notably Sir Karl Popper, who has declared that Darwinism is not a scientific theory at all, since it can never be proved or disproved: we cannot go back into the past to see what really happened. While conceding that it is a valuable framework for testing possible theories, Popper points out 'Neither Darwin, nor any Darwinian, has so far given an actual causal explanation of the adaptive evolution of any single organism or any single organ. All that has been shown – and this is very much – is that such an explanation might exist – that is to say, they [sic] are not logically impossible.'

Another kind of objection is the religious objection, but it is the objections of scientists which will here concern us. Scientists feel that, in the current theory, something is missing. 'These objections,' said Medawar, 'are very widely held among biologists generally, and we must on no account, I think, make light of them.'

The conference got straight to the point, opening with a paper by Murray Eden, Professor of Electrical Engineering at MIT, entitled: 'The Inadequacy of Neo-Darwinian Evolution as a Scientific Theory'. Eden proceeded to show that if it required a mere six mutations to bring about an adaptive change, this would occur by chance only once in a billion years – while, if two dozen genes were involved, it would require 10,000,000,000 years, which is much longer than the age of the earth.

Since evolution does occur and has occurred, something more than chance mutation must be involved. The role of mutation in evolution is a vexed question as we shall see. But plenty of other objections to Darwinism can be and have been raised.

In studying evolution, many scientists – paleontologists especially – have felt forced to accept the existence of some directive force and have felt it impossible to assign the many seemingly purposeful developments to chance. Here I will quote only one, though later we shall meet many others. The lot falls upon Richard Goldschmidt, formerly assistant director of the Kaiser Wilhelm Institute in Berlin and later Professor of Genetics at the University of California at Berkeley. 'I may challenge the adherents of the strictly Darwinian view . . . to try to explain the evolution of the following features by accumulation and selection of small mutants: hair in mammals, feathers in birds, segmentation of arthropods and vertebrates, the transformation of the gill arches in phylogeny including the aortic arches, muscles, nerves, etc.; further, teeth, shells of molluscs, ectoskeletons, compound eyes, blood circulation, alternation of generations, statocysts, ambulacral system of echinoderms, pedicellaria of the same, cnidocysts, poison apparatus of snakes, whalebone and finally primary chemical differences like haemoglobin versus haemocyanin, etc.' And he adds 'Corresponding examples from plants could be given.' We do not need to know exactly what all the terms he uses mean in order to see that he was deeply sceptical of natural selection. So puzzling did he find it that he was driven to reject the concept of genes as taught by current genetics. In this he was wrong – though today he does not look so wholly wrong as he did in 1940 – and this led to the neglect of his ideas, which are now being rediscovered. Twenty years ago, students were encouraged to snigger at the mention of his name. Today, however, many biologists are coming round to the view that he was pointing to the right problem.

Perhaps I will add one more instance, a short one this time. It comes from the geneticist Professor Richard Lewontin of Chicago University, who is baffled by the existence of so many different blood groups. Though these satisfy the simplest genetic assumptions, they 'are utterly mysterious in their causes and not a single case has been explained by the theory'.

While some scientists have felt forced to postulate some directive influence in evolution, others froth at the mouth at the mere idea.

This is because they fear that we shall revert to believing in a divine plan. Of course, if there is a divine plan everything can be accounted for simply by declaring it is part of the plan and a scientific approach becomes impossible. I shall argue, however, that we need not throw the baby out with the bathwater. I shall seek to show that a degree of directiveness is compatible with the body of scientific theory as it exists. After all, we know of one excellent example already. The fertilised egg is clearly programmed to produce an adult organism. The coordination of parts in a living creature is not achieved by chance but by precisely controlled development. We should not, *a priori*, exclude the possibility that there are programmes in evolutionary development too.

More recently, however, Darwinism has been under fire on different grounds; another issue has become the focus of attention, namely whether evolution took place gradually as Darwin insisted or by sudden leaps or steps. Evidence in favour of sudden changes was once known as saltationism (from the Latin *saltum*, a jump) but has reappeared as punctuationism.

Non-gradual theories had been advanced from time to time in the past. The issue was raised again in 1972 in a cautious paper by two American biologists, Niles Eldredge and Stephen Gould, entitled 'Punctuated Equilibria: an alternative to phyletic gradualism'. It was largely ignored: the time bomb did not go off for another five years, when they published a further paper in the respected journal *Paleobiology* entitled 'Punctuated Equilibria: the tempo and mode of evolution reconsidered'.

Eldredge had made a detailed study of trilobites, those creatures resembling large woodlice which were so successful some 500 million years ago, populating the primordial seas for hundreds of millions of years. In a long tour of the fossil beds of the American Mid-West, Eldridge noticed that their form did not change gradually with time; instead one version would drop out while a slightly different one would start up independently about the same time. He discovered five forms, each built on a distinctive plan, each of which remained stable while it lasted.

After a good deal of academic sniping, it was decided to call a conference on the matter and this was held at Chicago in 1980. 'Clashes of personality and academic sniping created palpable tension,' reported *Science*. 'We all went home with our heads spinning,' said one participant. Punctuationism, of course, was the explosive issue. Many of those attending suggested that the meeting was a turning point in the history of evolutionary theory. 'I know it sounds a little pompous,' Birmingham's Anthony Hallam told a journalist, 'but I think this conference will eventually be acknow-

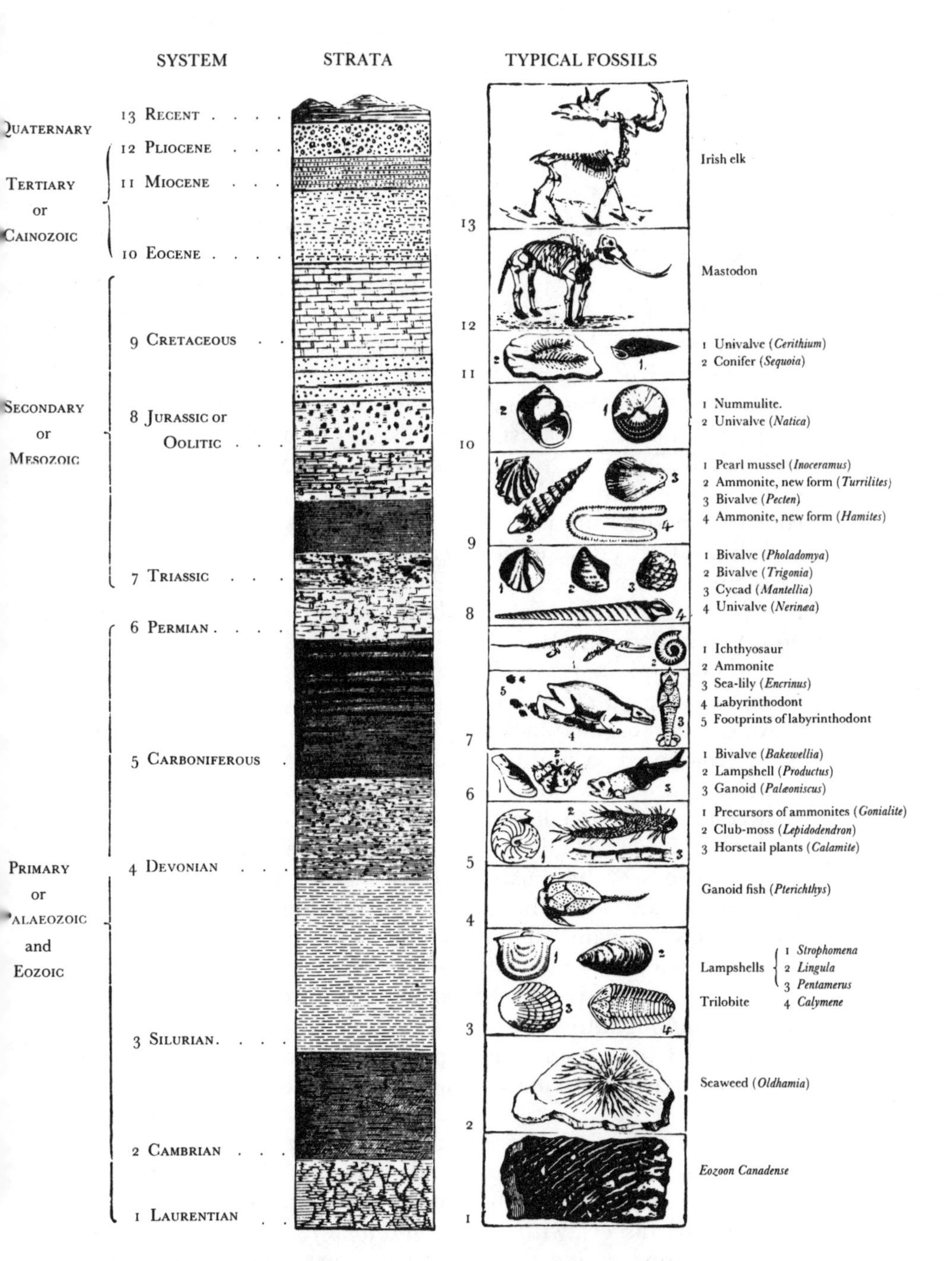

Table showing stratification of rocks through the various geological divisions, together with fossils typically found in each. To give an idea of the chronology, approximately sixty-five million years have elapsed since the beginning of the Tertiary period, two hundred and twenty-five since the start of the Mesozoic, and six hundred million over the scale as a whole

ledged as an historic event.' More cautiously, Gould just hoped it would lead to a rejig of ideas.

About the same time war erupted in Great Britain, when the Natural History Museum mounted an exhibition which drew the fury of (among others) Beverley Halstead of Reading University. The Museum had arranged its display on a basis that Halstead claimed was 'purely Marxist'. 'If it could be established that the pattern of evolution was a saltatory one after all then at long last the Marxists would indeed be able to claim that the theoretical basis of their approach was supported by scientific evidence,' he declared, in a heated letter to *Nature*. The basis of the display depended on an approach known as cladistics; if this was to be taken as received wisdom, he insisted, 'then a fundamentally Marxist view of the history of life will have been incorporated into a key element of the education of this country'.

'I think Halstead is completely mistaken' was the calm reply of the Museum's paleontologist Colin Patterson. Cladistics, he said, was merely a tool for studying the subject and has nothing to say about how evolution occurred. But Halstead cited Gould and Eldredge's article in *Paleobiology*, which contrasted the Marxist penchant for abrupt change with the Western preference for gradualism.

American museum officials watched the fracas with amusement, having mounted exhibitions based on the cladistic approach for years; but also with some envy, since the passions aroused at least showed that some people cared. As an ironical footnote, I might add that Gould is not a cladist and Eldredge is not a Marxist.

For decades evolutionists have concentrated on the differences between species. Now the emphasis is moving towards the larger issues – how the main evolutionary differences in body form emerged – termed macro-evolution. (Or by some, mega-evolution.) One of the many problems is the contrast between the enormous number of species, running into billions if we include all those which are now extinct, and the very few major divisions (or phyla) which have ever existed: not much more than thirty, depending on how you look at it.

The best known of course is our own: a body with four legs and a group of sense organs at one end. The same pattern serves for amphibians, birds, reptiles and most of the creatures with which we are familiar. Insects, with six legs and their main nerve cord on the ventral or under side, form another. Lacking a spine or bony structure, they depend on a hard outer casing for strength and rigidity. Fishes are different again, and radially symmetrical creatures like the jelly-fish are obviously even more distinct. The octopus represents another solution familiar to us. Most of the other patterns comprise obscure wormlike creatures dwelling, in many cases, beneath the water.

In addition to macro-evolution and micro-evolution, there is a third long-standing controversy, known as Lamarckism from the French biologist of that name. It proposes that alterations produced by circumstances, such as the swelling of muscles which accompanies repeated effort, or the hardening of the skin under pressure, can become inherited. In popular terms, it claims that the giraffe acquired its long neck by reaching up for leaves – though, as I shall explain in Chapter Two, this is not so, nor did Lamarck claim it was.

The Darwinists have claimed repeatedly that this doctrine is not only false but exploded, dead as a doornail. Unfortunately, it keeps raising its head.

As recently as 1980, a young Australian, Ted Steele, conducted experiments which he claimed gave support to the inheritance of an acquired characteristic – in this case immunological tolerance. Parallel experiments conducted by Leslie Brent at the Clinical Research Centre in North London did not confirm his work, but Steele claimed this was because they did not follow his method closely enough. When the two sides met in a radio debate organised by the BBC such violent argument broke out that the microphone was closed down while everyone recovered their cool.

However the case does not depend on whether Steele was right or wrong. In the thirties, Professor C. H. Waddington of Edinburgh conducted experiments which no one has challenged, which gave support to something very close to Lamarckism, although Waddington, fearing to be labelled as a crank, did not claim he had shown the acquisition of characteristics by a Lamarckian process but called it 'genetic assimilation'. Other experiments at plant breeding stations in Wales produced results equally hard to reconcile with the Darwinian paradigm.

These then are some of the issues we shall be examining in this book. But there is another, less obvious and of wider significance.

There are two senses in which an organism can be said to develop. It starts from a single cell – a fertilised egg or seed – to become a multi-cellular organism of great complexity – a process known to biologists as ontogeny. And it also develops over evolutionary time: we ourselves have descended not merely from a common ancestor with the apes, but also by a chain leading through fish and probably sea-urchins to some primordial single cell. (This biologists call phylogeny.)

Now these two processes must in some way be linked. If we doubted this our doubts would be resolved by the sudden appearance of atavisms, or structures from the past. The horse, for instance, once had three toes; occasionally modern horses display the three-toed form. Even more striking is the fact that in the womb we all, animals and men, pass through earlier phases in our evolutionary existence. For a few days, the human embryo develops gills. A rudimentary tail is formed and occasionally persists until birth and after.

But what is the nature of the connection? This remains an impenetrable mystery. We don't understand the forces which arrange cells in the complex patterns which form muscles, nerves and other tissues; nor how the tissues are organised into organs and limbs. Nor, if it comes to that, how the final integration into a functioning organism is achieved. The occasional birth of teratomas, of tissue masses in which can be detected partly formed limbs and organs, shows vividly that organisation is effected at more than one level.

Since we do not understand how the plan on which all these structures are built changes over evolutionary time, we cannot claim to have explained the evolutionary mechanism. If a fin becomes a leg, the moulding forces have palpably modified their plan – if they have a plan. Evolutionary history is a mass of such modifications; one could almost say it consists of them. Scales become feathers. Legs become wings. Stomachs become swim bladders. Even at the level of biochemical processes, substitutions and elaborations occur. All Darwinism has to say about such miracles is that they are due to chance.

Finally, there is the problem of behaviour. If a creature survives in the struggle for existence it is because it adopts a winning strategy. But how does behaviour evolve? Why are there such curious differences in the behaviour of closely related species, and do they matter? This too we shall have to discuss.

Many biologists, while ready to accept that neo-Darwinism explains minor changes in structure, changes of colour and so forth, are sceptical of its ability to explain those sweeping changes of plan such as the rise of terrestrial creatures from fishes, or of fishes from the spineless jellyfish and starfish which preceded them. This is one of the questions I shall explore very fully.

But the real objection to neo-Darwinism is that it 'explains' too much. (If explains is the word.) If a creature acquires a new feature, it can always be argued that it was advantageous. There are, for instance, a few genera of warm-blooded fishes. So it is argued that this enables them to swim faster and thus to capture fast-moving prey. But no experiments have been done, as far as I know, to show that this favours their survival. If it really is an advantage, why (one may ask) has it not become more general? Cold-blooded fish constitute an immensely successful group. As Professor G. C. Williams of Princeton has said: 'The mere presence of an adaptation is no argument for its necessity.' Neo-Darwinism, he complains, 'has provided very little guidance in the work of biologists'. It has provided few generalisations and is too often employed 'to give a vague aura of validity to conclusions on adaptive evolution'. And he adds acidly: 'A biologist can make any theory seem scientifically acceptable by merely adorning his arguments with the forms and symbols of natural selection.'

While evolutionists are quick to explain the new features which do appear they carefully avoid the complementary question of why some features fail to appear or appear so rarely. Why, for instance, are electric organs only found in fishes? And why are luminous organs absent in all freshwater fishes? Again, why are the little motile hairs known as cilia absent in arthropods and nematodes (spiders and eelworms are instances) while they are present in a wide range of other creatures ranging from protozoans up to man? And why are the pigmented cells which effect colour changes in creatures like the chameleon and in fish never found in warm-blooded animals? Surely it is just as advantageous for a warm-blooded animal to blend into its background as it is for a cold-blooded one?

A more recent convert to the ranks of the critics of neo-Darwinism is Professor T. H. Frazzetta of the University of Illinois, who observes: 'With each passing year the once rather simplistic views on evolution continue to crumble.' France's Professor J. P. Lehman puts it even more strongly: 'Darwinism in its ancient and classical form has broken down,' he declares.

But none of this means that evolutionary studies are at a standstill. Quite the contrary. 'Evolutionary biology is in an exciting transition,' asserts Stephen Stearns of Reed College, 'bubbling with controversies . . . A field full of fresh ideas and challenges to tradition, it should attract adventurous minds.' It is this sense of rediscovery, or imminent breakthrough, rather than one of carping criticism that I shall attempt to convey.

The fact of evolution is not in question. What is in question is how it occurred and whether natural selection explains more than a small part of it. As we shall see, a great many eminent biologists have raised this question but such has been the confidence and aggressiveness of the Old Guard that their views have been swept under the carpet and ignored. There has been a tightly knit school of neo-Darwinians, headed by such important figures as Ernst Mayr of Harvard, Theodosius Dobzhansky, most recently of Rockefeller University, George Gaylord Simpson, Agassiz Professor at the

Museum of Comparative Zoology at Harvard, in America; and in England the late Professors Julian Huxley and J. B. S. Haldane, who have trenchantly defended the orthodoxy, often in the face of the facts. So, today, nearly fifty years later, the time has come, it seemed to me, to face the facts and attempt what no one has yet done; namely, to lay out the evidence clearly and systematically.

CHAPTER ONE

What Darwin Said

1 Chance Or Purpose?

It was not by chance that living things, in all their limitless variety, appeared upon this earth. Nor was it by the working out of a preconceived plan of divine origin. Life seeks new and subtler forms of expression by a blind impulsion, an inner directedness. As water, falling on the hilltops, constantly seeks a lower level by every available channel and so finally forms streams and rivers, so life itself floods into every available opening, sometimes only to find its progress blocked, sometimes to plunge still further and to broaden out into pools and lakes of hitherto unexploited possibilities.

It was Charles Darwin's theory that evolution occurred by natural selection which made blind chance the arbiter of evolution. There were those before him, notably his uncle Erasmus Darwin and the French botanist the Chevalier de Lamarck, who saw that the Ladder of Nature was not a static system: that creatures could, as the generations passed, slowly climb the ladder. But it was Darwin who first had the temerity to suggest that this process was not the realisation of a divine plan. It was this, much more than the specific implication that men originated from apes or apelike stock, that so horrified many of his contemporaries.

But today, a century and a half since Darwin's conception was born, it is becoming increasingly doubtful whether natural selection explains more than a part of the evolutionary story. It accounts brilliantly for the minor adaptations which living organisms make to meet the challenges of the environment but it is by no means clear that it explains the major changes in evolution: the change from spineless jellyfish to fish with brains and backbones, for example, or the change from fish to air-breathing, four-legged land animals, to name only the most obvious examples.

And there are other things it does not explain. Just to take a preliminary example, consider the almost incredible behaviour of the planarian worm known because of its small mouth as *Microstomum*.

There is a small creature known as *Hydra* which you may have met during biology lessons. It is about the length of a capital 'I' in this type and it clings to underwater plants. It consists of little more than a tube, with a mouth surrounded by waving arms which direct food into it and a foot by which it attaches itself. Its most engaging habit is to proceed by somersaulting. It bends its head over to the surface on which it is posed and then detaches its foot which it then puts down in a new position.

Some species of *Hydra* develop stinging cells known as nematocysts. Each of these cells contains a coiled, poisoned hair which can be ejected with explosive force. Another sensory hair projects outside the cell and serves as a trigger, discharging the nematocyst as soon as it touches anything. These stinging cells are arranged in batteries on its surface.

This is curious enough, but stranger by far is the way in which the planarian worm known as *Microstomum* has exploited this mechanism. It has developed a ploy which strikes at the heart of evolutionary theory and indeed defies explanation on any grounds yet available to science. This species varies its normal diet by eating the *Hydra* in question. But it does not digest the nematocysts or the immature cells which give rise to them. Somehow it passes them through its body and positions them on its surface – or skin, so to say – with the stinging points outwards. Then, when enemies approach, it discharges these nematocysts; and in one variety it does not even wait for the assailant to touch it – it discharges the poisoned darts at it like shells or rockets.

When fully armed, *Microstomum* ceases to feed upon *Hydra* and returns to its usual diet. But after it has discharged its weapons it makes a new meal of *Hydra* in order to rearm itself. In order to carry out this extraordinary programme, three different kinds of tissue within *Microstomum* must cooperate: endoderm, parenchyma and epidermis, to be precise. How has it acquired this complex routine? How too has *Microstomum* learned to regulate its diet? All this in a creature which has no brain or nervous system. Yet it implies a memory and an inherited instinctive pattern.

The theory of evolution by natural selection is powerless to explain how chance variation could have evoked such a closely coordinated programme. Nor can physiologists explain how the cells from the

Hydra migrate through the body of the *Microstomum*, although rather similar migrations occur when the fertilised egg of a mammal begins to develop into an embryo. For instance, cells from various parts of the cell mass come together in one place to form the heart, as if they knew it was their destiny to become 'heart' and nothing else. Here again we seem to see a purposiveness of the kind which Darwinists refuse to believe in.

If Darwin plunged us into a purposeless world of chance it was because he was in reaction from a philosophical and theological position which not only ruled out scientific enquiry but denied man's freedom to modify his own future. But is an evolution based on chance the only alternative to one based on a divine plan? The story I am going to tell will show that it is not.

And be that as it may, the evidence is accumulating that chance is insufficient to explain the appearance of marvellously coordinated structures and perfectly adapted behaviour which biologists are uncovering. As two American biologists have recently declared: 'The main puzzle, the problem of the mechanism of evolution, is far from being solved.'

In short, the dogma which has dominated most biological thinking for more than a century is collapsing. This marks a turning point in the history of science, one of those 'paradigm shifts' (to use a current term) which transform our world view, just as the notion that the sun, not the earth, was the centre of the solar system did.

2 *Eminently Curious*

It was in 1835 that the young Charles Darwin – he was twenty-six – reached the Galapagos Islands and there noticed a number of finch-like birds. His diary does not mention them and there is only a brief mention in *The Voyage of the Beagle*, but these birds were to change the course of biology.

The Galapagos Islands lie some 600 miles west of South America and the scenery resembles, Darwin said, what one might imagine 'the cultivated parts of the infernal regions' to be like. 'The inhabitants of these islands,' he wrote in his diary, 'are eminently curious.' He was thinking, no doubt, chiefly of the giant lizards and tortoises, for the finches are drab in appearance and feeble in song. It was not until after Darwin's return home that the ornithologist John Gould identified these birds as related to the finches. There were thirteen species on the Galapagos with another on Cocos Island to the north. What Darwin was acute enough to notice was that they had developed four distinct modes of existence. Six species lived on the

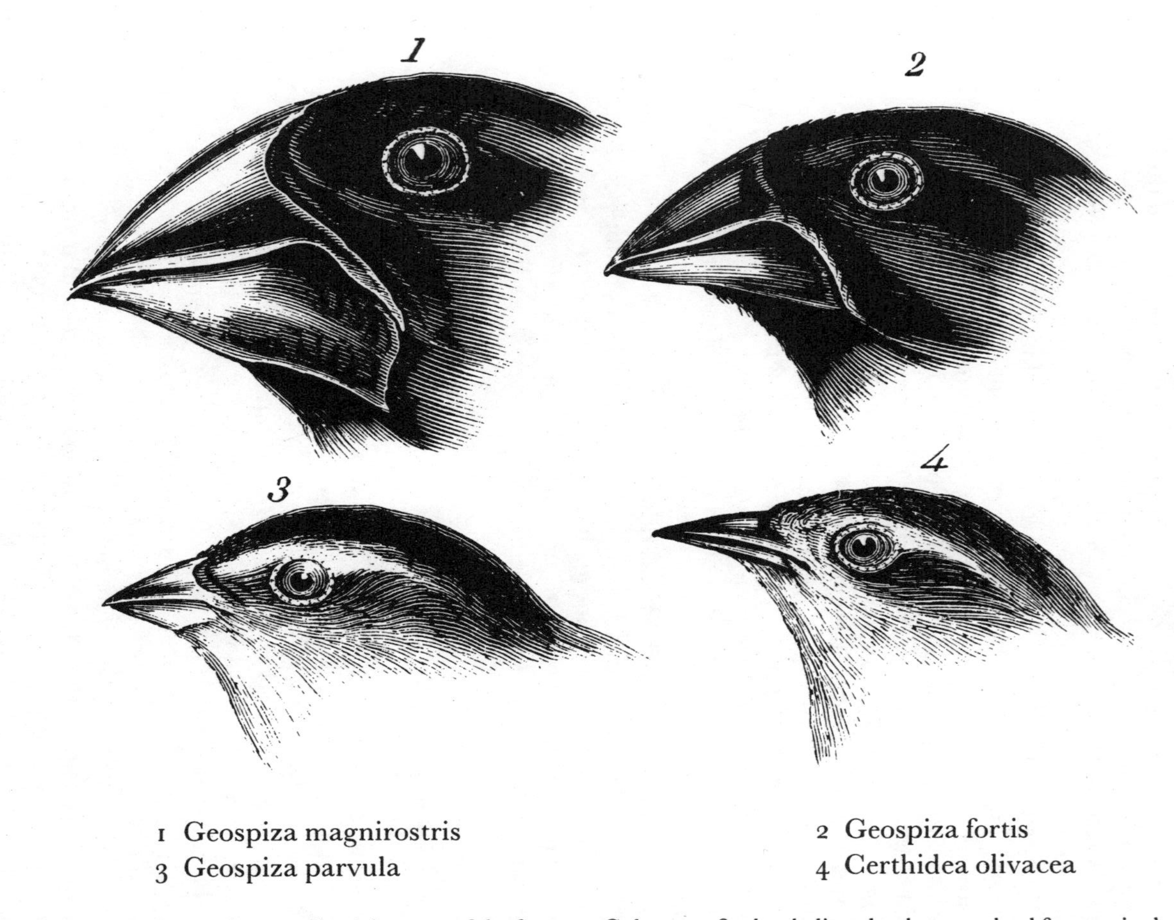

1 Geospiza magnirostris
2 Geospiza fortis
3 Geospiza parvula
4 Certhidea olivacea

Example showing different shapes of beak in some of the fourteen Galapagos finches believed to have evolved from a single species

ground in coastal areas. Another six lived in forest trees. One more species lived in bushes, while the Cocos Island species lived in tropical forest.

Of the six species of tree-finches, one species was vegetarian and lived on buds and fruit; three lived on insects; another lived in mangrove swamps and ate the insects there; while the sixth proved to be one of the world's most extraordinary birds. As does a woodpecker, it looked for insects in the bark of trees, using its chisel-like beak to expose them. But, unlike a woodpecker, it did not have a long tongue to fish them out with, so it took a twig or spine in its beak and probed with this – the only known instance of a tool-using bird. The use of tools by animals is exceedingly rare, except among the higher apes.

Darwin noticed that the beaks of these birds were slightly different in each species, depending on their diet. The ground-living species likewise had beaks adapted to different sizes of seeds, while the fourth, which lived on prickly pear, had a longer, more pointed beak.

The Galapagos were too far from the mainland for it to be likely that different species had arrived on different occasions, so Darwin felt safe in assuming that all fourteen species had evolved from a single original type. (How *it* had got to the Galapagos remains a mystery.) Cogitating on this fact led him to the conclusion that species arise from the slow accumulation of slight structural differences – differences which bestow some advantage on the possessor, increasing its chances of surviving to maturity and leaving offspring. This, in essence, was the theory of 'natural selection'.

Darwin's book was entitled: *On the Origin of Species by Natural Selection or the Preservation of Favoured Races in the Struggle for Life*, and in it he summarised his theory in these words: 'As many more individuals of each species are born than can possibly survive; and as, consequently, there is a frequently recurring struggle for existence, it follows that any being, if it vary however slightly in a manner profitable to itself, under the complex and sometimes varying conditions of life, will have a better chance of surviving and thus be *naturally selected.* From the strong principles of inheritance, any selected variety will tend to propagate its new and modified form.'

Actually it was only in March 1837 that, as a result of conversations with Gould, Darwin became an evolutionist, and not until September 1838 that, after reading Malthus on population, he had the great moment of illumination. 'Here at last I had got a theory by which to work,' he wrote later.

If Darwin had confined himself to saying that advantageous changes would be perpetuated by natural selection, there would be nothing to worry about. But he went much further. He said (1) that *all* changes which became fixed did so in this way; (2) that all changes occurred by *imperceptible gradations*; and (3) that all changes arose in the first instance by *chance*. The second of these propositions is certainly untrue. For instance, the eye-colour of fruit-flies mutates from red to white, which is by no means imperceptible. The first proposition has recently been challenged by the 'neutralists', led by Professor Motoo Kimura of Japan, who show that changes which are neither advantageous nor disadvantageous can become fixed in inheritance. The third proposition excludes the possibility that changes arise in response to needs, which we shall look at in the next chapter. This is the view known as Lamarckism.

Although he called his seminal work *The Origin of Species*, Darwin pushed his idea further, claiming that not merely the difference between species but that all such differences – those between birds and mammals, between fish and insects, between corals and reptiles – had arisen by the same process. If, way back in the past, a primitive creature diverged into two forms, each of these might in turn diverge, and so on, leading to quite different outcomes. Here I think I must digress for a moment to explain the terms employed by biologists for the purposes of classification. They begin by distinguishing three 'kingdoms' which are radically different: plants, animals and fungi. (Plants use sunlight to make nutriment from simple chemical substances. Fungi cannot do this and depend on plants. Animals depend on plants too but, unlike fungi, they can move about in search of food.)

The kingdoms are divided into major groups known as phyla (singular: phylum). For instance, all vertebrates constitute a phylum. Creatures with jointed legs and an external skeleton, such as crabs and insects, constitute the phylum known as Arthropoda. Phyla are divided into classes. Thus the vertebrate phylum contains four classes: birds, fishes, mammals and reptiles. Classes are divided into orders, orders into families, families into genera, genera into species. Thus the tiger and the lion are two species within the genus *Felis*. As this does not always work out neatly in practice, taxonomists are sometimes forced to postulate sub-orders, sub-species and even varieties. In short, we – men and women – are members of the phylum Vertebrata, the class Mammalia, the order Primates, the sub-order Anthropoidae, the genus *Homo* and the species *sapiens*.

Unfortunately taxonomists often disagree whether a particular group should be regarded as a family, a sub-family or a genus, or whether a group of species should be treated as a single family, or genus, or as several. For example, some taxonomists treat all the cats, except the cheetah, as a single genus *Felis*, while others distinguish the larger cats, like lions, and the bob-tailed cats, like the lynx, from the smaller cats. A convenient general term for a taxonomic group, whatever its rank, is taxon (plural: taxa).

We shall not be much concerned with these taxonomic distinctions, except for the major distinctions between phyla. There have been thirty-one phyla in evolutionary history, of which nine are now totally extinct. Many biologists, while ready to accept that natural selection could give rise to species and even genera, have found it difficult to believe that it could account for the vast differences between phyla. A fish, for instance, is so very unlike an insect that something much more than the accumulation of slight differences seems necessary to account for the contrast. It is not a question of a slight competitive advantage; they lead completely different lives.

Another question which Darwin failed to ask himself – understandably enough, since he was in the Galapagos only for a short time and had much else to observe – was whether there were any intermediate forms between the species; and if not, how this was to be accounted for in the light of his theory.

It was not for nearly a century that the question was to be posed, and this time not in the Galapagos but in the Pacific. Islands, of course, are a boon to evolutionists, since, broadly speaking, the species on them derive from single creatures arriving at long intervals, so that the evolution story is relatively unconfused. The Hawaiian group of islands was formed only some three million years ago, the islands at the north of the chain emerging first. Hawaii itself is only one and a half million years old, so the time in which diversification occurred is more narrowly defined than is usual.

The fauna and flora of the group are very varied: the place is a biologist's paradise. Back in 1906 Thomas Guppy, a missionary who was also an evolutionist, went there to study, among other things, the honey-creepers, birds known to taxonomists as the Drepanididae.

'When the ancestral pair of drepanids first arrived, and found this land rich with beetles and various other insects, and gay all the year round with lobelias and other honey-bearing flowers, yet with no rival or enemy in sight, it had before it an evolutionary opportunity such as can scarcely have been duplicated in the whole history of avian life. The standardising effect of close natural selection was removed. Food was very abundant but probably differed in one way or another from what those birds had formerly been used to, so that they may even from the start have begun to turn aside from their previous habits of feeding.' The words are those of a later naturalist, A. Gulick. He goes on to describe how the birds' beaks began to vary in form, according to their diet, just as in the case of the Galapagos finches. Some ate insects, some hard sandal nuts, some fruit. Some picked insects from the bark, some ate seeds, some looked for insects.

And some developed a highly novel preference: they drank nectar from flowers.

In short a family once homogeneous is now divided into eighteen genera, 'ranging from stocky, seed-eating birds with bills like cross-beaks and parrots, through finch-like birds that glean smaller seeds, to little creatures with long, thin, flexible bills for gathering honey and insects out of tubular lobelias, and even birds with a short stout lower mandible and a very long slender upper one, the first usable for pecking away loose bark, and the other for probing out the grubs of the native boring beetles. Adaptation has become completely and narrowly specialised for feeding upon the nectar, seeds and insects native to their Hawaii.'

The German biologist Richard Goldschmidt makes the following comment: 'It is well worth while to enquire whether such an evolution can possibly have taken place on Darwinian lines. How the woodpecker or stonecrusher type, which involves, of course, the whole anatomy, may have evolved from a honey-sucker type by a series of micro-mutations controlled by selection is simply unimaginable, and one can understand why so many geneticists stick to the Lamarckian explanation.' And he adds 'We cannot understand, either, why all these different lines of evolution should have blossomed out simultaneously, even if a neo-Darwinian interpretation of the resulting type were feasible.'

3 Apparent Objections

Darwin's early observations on finches raised an absolutely crucial question and one which is central to the theme of this book. It is this. Did the finches turn to different foods because their beaks were becoming different, or did the beaks become different because they had chosen to eat different food? If Darwin's theory is correct, the latter must be the case. It would be of no advantage if a bird's beak changed so as to be adapted to eating fruit when its habit was to eat insects; it would be an actual disadvantage. But in that case, why did some of the finches decide consistently to consume a different food from the others? It must be borne in mind that there is no shortage either of insects or of fruit on the Galapagos. And even if there had, at some past time, been a shortage of one, why did not all the finches change their regimen? In a word, it seems that it was the *behaviour* of the finches which brought about, via selection, the structural change. And if we needed confirmation, there is that unique tool-using bird, which had changed its diet even though evolution had *not* provided the means. Was ever a hint more clearly given? The same problem arises even more pointedly in the case of the honey-creepers.

In this book, I shall offer you many further instances where a change of behaviour seems to be the critical factor in evolution and I shall eventually consider the evolution of behaviour itself. Here it is simply interesting to note that, right at the beginning of the story, an element of doubt creeps in.

After his return to England, Darwin explored the topic in much detail, breeding earthworms, talking to fowl-breeders, and much else. After drafting his book, with typical caution he consulted a number of eminent naturalists and was advised not to publish it. So he put the manuscript away and spent eight years preparing a monograph on barnacles. It was a very thorough job, and it was for this that he was admitted as a Fellow of the Royal Society, and not for his wild evolutionary theories, as the Society was careful to explain. Then the news came that Alfred Wallace had had the same idea and was about to publish it. Darwin dusted off his manuscript and the two versions were given to the world simultaneously.

While some of those who advised against publication originally may have felt that the work would cause a scandal, others must have seen how weak it looked, given the state of knowledge at the time. Three things in particular – as Darwin fully realised – made the hypothesis doubtful. First, he had no idea what, if anything, could cause the variations which natural selection was expected to select. Why did not the finches hand on the standard beak-pattern to their offspring indefinitely? Second, Darwin himself believed that when two creatures mated, the hereditary endowment was simply mixed together and a portion bestowed on the offspring. Thus a new character arising in one parent would be halved in the first filial generation, reduced to a quarter in the second and would soon be diluted to insignificance. This was fatal to his theory. Only if there were hereditary units (genes as we now say) which would either be bestowed or not bestowed, would some offspring have an advantage over others and be more likely to survive.

The belief in 'blending inheritance' is common and the appearance of offspring resembling in part each of their parents seems to support it. This occurs because most of the obvious characteristics result from the interaction of several genes. But Darwin should not have been taken in by this since in 1802 a member of the Royal Horticultural Society had written to point out that when he bred together green unwrinkled peas with yellow wrinkled ones, the next generation consisted of four varieties: green and smooth, green and wrinkled, yellow and smooth, yellow and wrinkled. In short, the characters *green* and *smooth* were either transmitted or not transmitted. As biologists say, they were segregated. Furthermore, Thomas Knight, a very well-known botanist, a Fellow of the Royal Society and recipient of the Copley Medal, had written papers to much the same effect when Darwin was a young man.

Thirdly – and it was a question which worried Darwin very much – was there enough time for the accumulation of these minute changes to produce such large effects? Geologists of the period, reacting from the absurd 6,000 years claimed by theologians, were inclined to give the earth almost unlimited age, but from 1862 the great physicist Lord Kelvin argued that the earth was probably less than a hundred million years old. His case rested on three arguments, of which the main one was that the earth would have cooled off to its present temperature in about that time or certainly not more than 400 million years. He did not know, naturally, of the existence of radioactivity which produces heat, and that the earth's core contains radioactive materials which restrict its rate of cooling. However, as scientists so often do, he announced his conclusions with the utmost confidence and left Darwin greatly shaken. Kelvin himself declined to believe Darwin's theory to the end of his life.

Little wonder, then, that Darwin's scientific acquaintances advised him not to publish his implausible theory.

But in due course the pendulum was to swing to the other extreme.

Even while the Darwinian controversy was at its height, the Austrian monk, the Abbé Mendel, published in an obscure journal his two short papers concerning inheritance. He showed, after eight years' work, that in peas at least characteristics were segregated. In his main experimental series, he obtained 8,023 peas from 258 plants. Of these 6,022 were yellow, 2,001 were green, a ratio of three to one. The same ratio appeared in the seven other factors he studied. From this he made the acute deduction that the yellow character dominated the green. Yellow-yielding plants had either received a factor for yellow from each parent, or a factor for yellow from one and green from the other. Only plants receiving factors for green from both parents would appear green. Such factors are called dominant and recessive, respectively.

Not until the turn of the century was this work brought to general attention. For Darwinians the concept of discrete inheritable factors – genes – was pure gold. Gone was the spectre of 'blending inheritance'. Stable hereditary units were just what was needed to preserve advantageous variations.

Of course, the idea of genes was strenuously opposed by the old guard. Genes had no objective existence, they said; they were 'just concepts'. Or if they existed, it was only in a few species. As late as 1908, the biometrician Karl Pearson declared: 'There is no definite proof of Mendelism applying to any living form.' With other opponents, he actually founded the journal *Biometrika* to campaign against Mendelism. However, Mendel turned out to be right.

Then the problem of the source of variations was solved – or so it seemed at the time.

In 1900, Hugo de Vries, a Dutch professor of botany, brought out the book in which he put forward his mutation theory. In 1886 de Vries had found some unusual dwarf and broad-leaved forms of the evening primrose, *Oenothera Lamarckiana*, growing, together with the normal form, on a piece of waste land near Amsterdam. The evening primrose was an import from America and these plants had seeded themselves from a neighbouring garden. De Vries rented a house nearby and proceeded to breed the abnormal form. He found that in the next few generations many variants arose and that these seemed to breed true. To de Vries it seemed that the species was disintegrating into forms so distinct as to constitute new species. Pursuing this exciting lead, in the next ten years he raised 53,509 plants and found among them what he considered to be eight new species.

Even before this discovery, de Vries had become dubious about the Darwinian view that species are modified by the accumulation of many slight changes, and had begun to look for sudden jumps or 'saltations'. (Two earlier biologists had already suggested that something of the sort occurred.) Then in 1894 William Bateson, a leading British geneticist, published his classic work, *Materials for the Study of Variation*, which described many other examples. In his book de Vries further annoyed the Darwinians by saying that one ought not to ask: what is the origin of a species, but rather, what is the origin of a species character?

As it turned out, what de Vries had painstakingly gathered were not new species at all, but stable intermediate types. But the idea of mutation proved sound, and many experiments confirmed that jumps do occur.

It was the American T. H. Morgan, who had started out as an embryologist, who at the age of forty-three suddenly realised that mutation as proposed by de Vries could provide a brand-new way of studying genetics. Ironically enough – the history of science is full of ironies – he had at first been highly sceptical of Mendelism and had agreed with Pearson's comment.

Morgan soon showed there were proofs. After experimenting with rats, mice, pigeons and even aphids and plant lice, he finally settled on the fruit-fly, *Drosophila melanogaster*, as suitable material. It attains maturity in only twelve days and thirty generations can be bred in a year. By the end of 1910, Morgan had found more than forty mutant forms: some were wingless, others hairless, eyeless or of unusual colour, and so on. Here, it seemed, was the source of variation for which Darwin had been looking for so long.

Meanwhile, the physicists had discovered radioactivity and had even begun to evolve methods of dating ancient rocks by radioactivity. These techniques showed the earth to be at least three billion years old, a figure later raised to four and a half billion. Darwin's worst obstacles had melted away. Finally, in the 'twenties, Professor J. B. S. Haldane and the statistician Sir Ronald Fisher provided the whole structure with a mathematical foundation. They pointed out that it was populations which evolved, rather than individuals, so that one could calculate the frequency of mutation in the population as a whole, even though what would happen in single individuals was unpredictable. Natural selection theory now hardened into a dogma known as neo-Darwinism. In schools it was taught as a known fact. A few dissentient voices were raised, but facts which did not fit the theory were brushed under the carpet. They would be cleared up some day or other.

Scientists have tidy minds and like to think that the complexities of nature can be reduced to simple principles. I am reminded of the position in nuclear physics half a century ago, when the atom was thought to be composed of protons and electrons only, so that the whole material universe could be accounted for in terms of just two oppositely charged particles. Then this admirable simplicity began to melt away. Neutrons began to complicate the picture, then mesons, until today we recognise something like a hundred subatomic particles.

The tidy simplicity of Darwinian theory is likewise beginning to crumble. As we shall later see, the genetic aspect has already done so.

4 *Puzzling Features*

Long before I began to research the subject in any detail, I had brooded about a number of puzzling features – things which didn't seem to fit the argument – which the textbooks largely ignored.

There is, for example, the fact that some creatures fail to evolve but chunter on quite successfully as 'living fossils.' Bees preserved in amber from the Tertiary period are almost identical with living bees. And everyone has heard of the coelacanth, supposed to have been extinct since the beginning of the Cretaceous period, 130 million years ago. But in the winter of 1938–9 a trawler working off the coast of South Africa dredged up a living coelacanth. And in 1952 several more specimens were recovered. The plant world also offers living

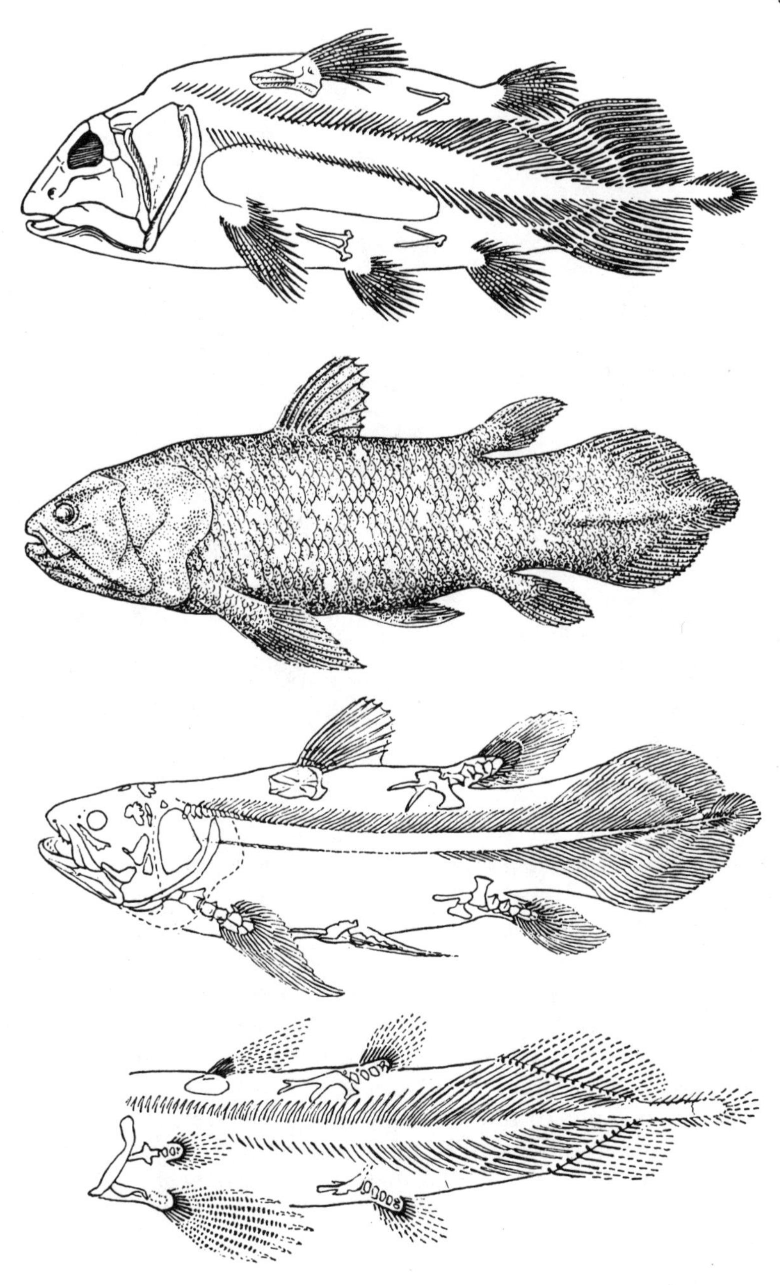

The coelacanth has displayed little evolutionary change for many millions of years. From top to bottom: fossil of an upper Jurassic specimen; *Latimeria chalumnae*, the only extant species; and details from the skeletons of *L. chalumnae* and a Jurassic coelacanth

Leaves and seeds of *Gingko biloba*, a 'living fossil' among plants

fossils, such as the gingko, with a leaf unlike that of any modern tree.

The greatest advance in the evolution of fishes was from a form without bony jaws to a more efficient form equipped with jaws. 'The importance of this evolutionary development can hardly be overestimated', writes the former Curator of Fossil Reptiles and Amphibians at the American Museum of Natural History, Edwin H. Colbert. It 'offered new possibilities of evolutionary advancement that expanded immeasurably the potentialities of these animals'. Why, then, do we still find lampreys, which are jawless fishes, doing very well today? If possessing jaws was such a wonderful advantage, why did not the jawless fishes realise how backward they were and succumb?

These primitive vertebrates take us back to 500 million years ago; but a still more extraordinary example of failure to evolve is found in the bacteria. Since they reproduce themselves, in favourable conditions, every twenty minutes, they might be expected to evolve faster than any other organism – but fossil bacteria going back three and a half billion years, to the threshhold of life itself, have been recovered and are virtually identical with modern forms.

There are really two problems here. First, why did some species fail to give rise to superior forms? Second, why, when they did give rise to superior forms, did not the inferior forms die out, worsted in the evolutionary struggle?

In contrast with these static species are those in which an evolutionary trend seems to continue far past the point where it is advantageous to a stage where it seems positively damaging. The classic example here is the Irish elk, *Cervas megaceras*, which is neither Irish nor an elk but is a Siberian deer. Its antlers were gigantic: twelve feet across and weighing nearly a quarter of a ton. As the Cambridge zoologist G. S. Carter writes: 'But the enormous size of the antlers of the elk makes it difficult to believe that they were evolved by the processes of natural selection. At this size they must have interfered seriously with the activity of the animal, and therefore, one would have thought, have been disadvantageous and removed by selection.' One would, indeed.

Walter Modell, who has made the study of antlers his special interest, says: 'The antler is a strange and uneconomic experiment, extremely costly to its possessors in many ways, and it seems destined eventually to disappear.' Which suggests that the real mystery is why antlers appeared at all, leave alone why they sometimes became so huge.

The official explanation for the gigantic horns of the elk is advanced by Professor George Gaylord Simpson. Horn size is correlated with body size, he maintains. In combat with other males for access to females, the bigger animal will usually win. Horns are secondary sexual characteristics and are used in the combat. Consequently, large horns are preserved. But the fact is, when deer fight seriously they fight with their feet. In antler fights, the antlers often become locked, and both animals die. Antlers are thus a disadvantage to the species. However, the elk died out and I cannot avoid concluding that a trend, originally useful, has in such cases gone too far.

In truth, increase of size has often gone too far, as in the well-known case of the dinosaurs.

Equally impractical is the long, dangling tail of the lyre-bird, which must handicap it in walking as well as in flight. It was evolved for sexual display. The lyre-bird builds a mound of earth on which it stands and sings, while bringing its remarkable tail, composed of specialised feathers, over its head and making the filaments in it vibrate so fast they look like shimmering light. But the question at once occurs to me: why does the female lyre-bird demand such a

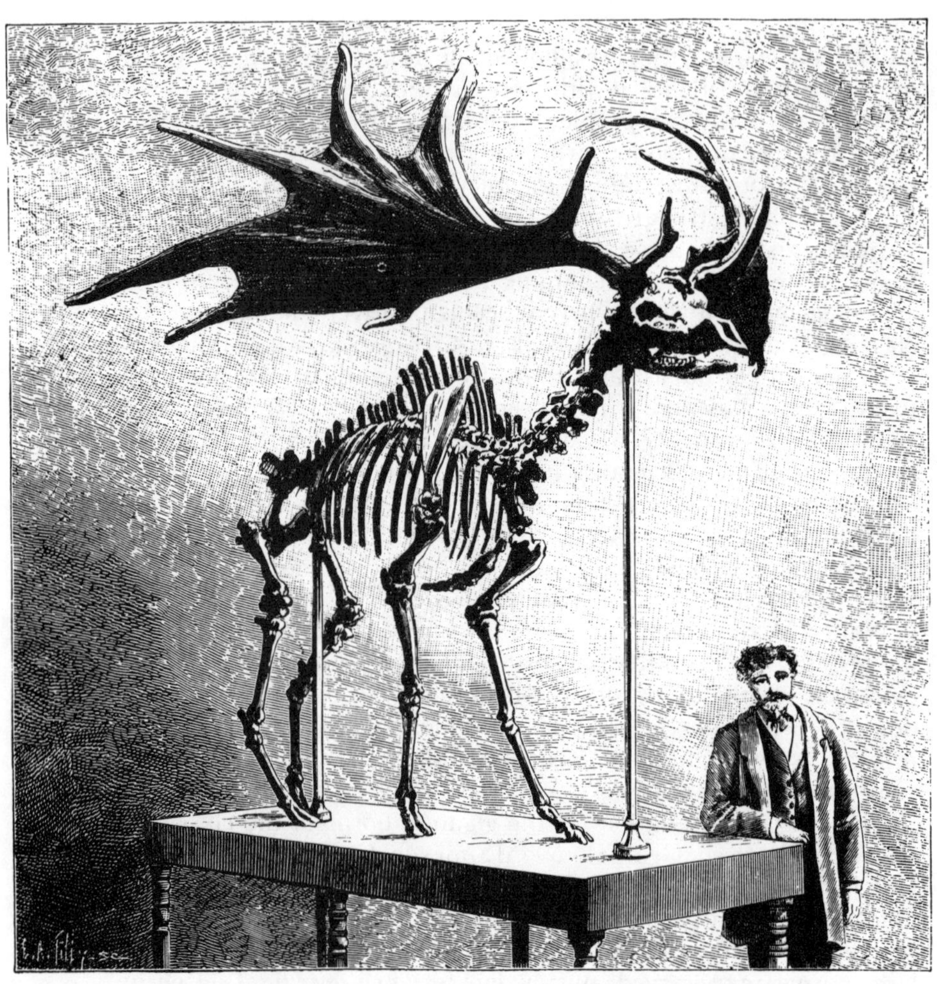

Skeleton of the extinct Irish elk, showing disproportionately huge antlers

pinnacle of virtuosity? Presumably she is going to settle for whatever is going. Or, if not, why have not other birds pushed their courtship behaviour to similar extremes? One bird which has, of course, is the peacock. But here again it is difficult to think that the peahen, as well as being dazzled by the peacock's ostentation, goes so far as to demand that the tail be covered with ocelli, as the eye-like patterns are called. The artistic complexity of the design seems more like a creative whim, and evolutionists do not attempt to explain it.

Often these enlargements proceed to extremes which defeat the original purpose, like the tusks of the mastodon which curved round so much they could serve no offensive function but rather acted as fenders.

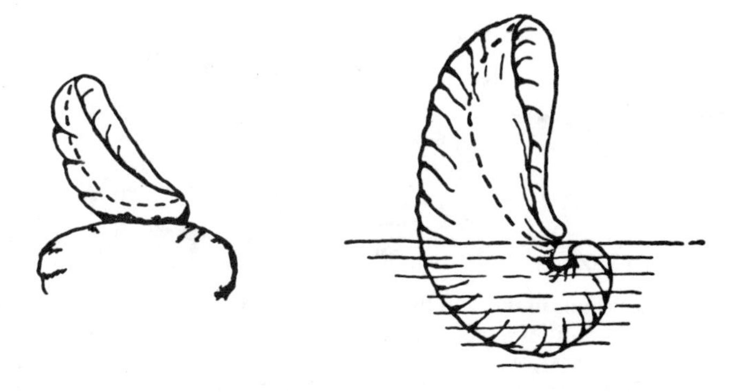

Stages in the evolution of the extinct oyster *Gryphaea*. Earlier forms were attached to another object. The area to the right of the broken line comprised the living chamber

I think it is obvious that a trend which was initially useful has continued to an extreme, and I shall refer to this as 'overshoot'. Some writers have described this as 'momentum' but the analogy is misleading. It is more like the case of a jammed switch, stuck in the 'on' position.

In any case, overshoots of the kind I have just described are not confined to size or the elaboration of display. A case which may be harder to explain away is presented by a type of oyster known as *Gryphaea*. The shell of this mollusc began to coil early in its evolutionary history and this coiling was closely studied in the 1920s by A. E. Trueman, who took advantage of the fact that an unusually long series of fossil oysters was available. In the earliest forms the coiling amounts to a mere ten degrees or so but it steadily increases and culminates in an impressive 540 degrees, or one and a half turns. At this point the oyster begins to have difficulty in opening its valves and coiling cannot proceed further without it perishing of hunger.

Desperate for a rationalisation, evolutionists suggest that the coiling of the creature's body raised *Gryphaea* out of the mud in which it lived, though whether this was really advantageous is anyone's guess – considering how many oysters manage very well without it. Nor is *Gryphaea* an unique instance. Thus there are molluscs, living on coral reefs, the shells of which have become so thickened that they can barely open enough to admit the plankton on which they feed. The only reasonable view is that evolutionary trends sometimes acquire a momentum which carries them forward long past the point at which natural selection should have eliminated them.

A third obstacle to a belief in the universality of natural selection can be found in those variations of form which seem quite random. As you may have noticed, some plants have leaves arranged in pairs,

Inherited variations in the fruit fly, *Drosophila*. In all but the male and female on the left these abnormalities have been produced by human intervention

one on either side of the stalk, while others prefer to place their leaves alternately. Again, you may have observed that in some leaves the veins run parallel with one another while in other species they form a network. Moreover, there are no intermediate forms. In such cases, does one form offer a selective advantage over the other? There is no evidence that such is the case and plants of both types flourish happily in the same environment. Look closely at the turf of downland and there you will find a score of flowers, grasses and mosses within inches of one another – quite different in form yet occupying the same habitat. In what sense are they all 'adapted' to it? There will be flowers with four petals or with five or six or even eight. Is there any advantage in having five petals rather than four? It does not seem so. There will be flowers with one stamen or several; leaves which are smoothly oval and leaves cut into lacy shapes. Is there any advantage in having a few additional hairs at the mouth of the corolla? We are forced to admit that some variations of form, at least, are quite random. That is, they are neutral as far as natural selection is concerned. So how did they arise and become stabilised?

Today biologists, who once scorned the idea of randomness, are readier to concede the point and even accept that the genetic constitution of a group of organisms can 'drift' with the passage of time as a result of these chance factors.

There are other indigestible observations which cannot easily be fitted into the Darwinian account. For instance on one of the Hawaiian islands there are no fewer than 300 species of *Drosophila* as compared with half a dozen or so on neighbouring islands. What this suggests to me is an abrupt genetic change – quite different from the accumulation of minute variations postulated by Darwin. These species differ not because there was any great advantage in differing but because of a genetic crisis. And as Richard Lewontin, another of America's leading geneticists, has despairingly said: 'We know virtually nothing about the genetic changes that occur in species formation.'

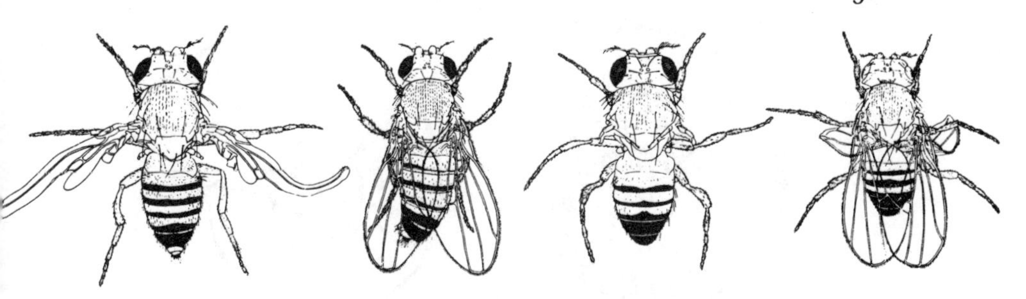

There are of course instances of this in the animal world as well. In your youth you probably hunted in quarries for ammonites, those decorative spiral fossilised organisms whose origins go back to Cambrian times at least. They bear ridges at the point where each compartment (for internally they reveal a series of compartments) abuts the next. Early in evolution only the first formed compartments show such ridges but with the passage of time they become more numerous and more elaborate. There is a definite trend, which suggests that some directive force was at work, but it is hard to believe that the ridges offered an evolutionary advantage.

Let me mention one more bothersome point which must occur to anyone considering, however superficially, the doctrine of natural selection.

Darwin assumed that the members of a species would be in competition with one another, and the species with other species. The animal which ran faster to escape its pursuer or which fought back more effectively, or which excelled in some other relevant way, would be more likely to survive than its rivals and thus to leave offspring carrying the genes which bestowed this advantage. But the assumption of universal competition is ill founded. Darwin's own example was of wolves pursuing a deer; he thought the fastest wolf would effect the kill and get the meat. But the fact is wolves hunt in packs and share the kill.

In fact competition is rather rare and animals evolve many mechanisms to avoid it, such as the defining of territories from which possible rivals are excluded. Another method is specialisation of diet, and Darwin's finches themselves provide an instance. The finch that lived on insects was not in competition with the finch that lived on a vegetarian diet, and so on. Was this then the origin of their subdivision into several species? It is a more plausible explanation than the Darwinian one, for there were plenty of buds and insects to spare, so it is unlikely that they specialised in diet because of a shortage of food. In a very similar way, as Konrad Lorenz, the Austrian

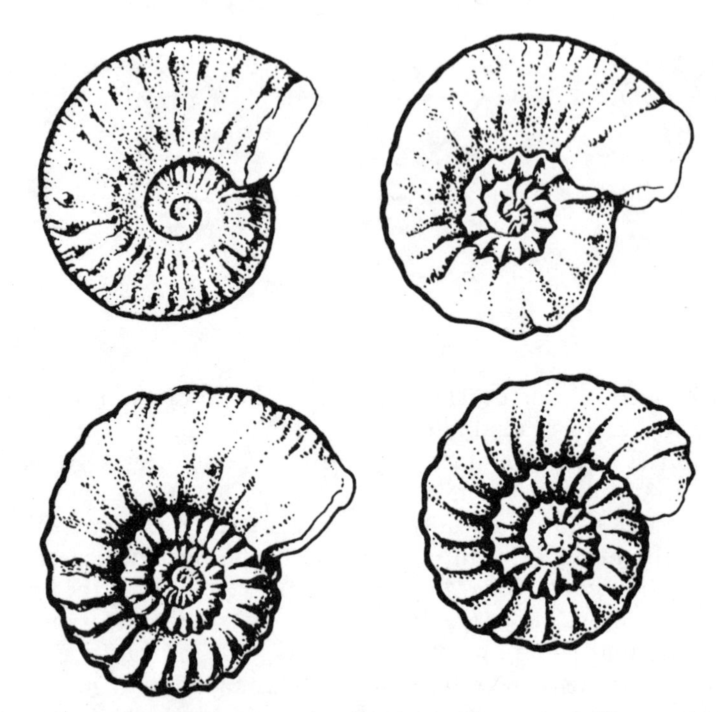

Successive species of ammonites, showing the development of ridges on the shell. From the top, left to right: *Liparoceras cheltiense* has no ribs, *Androgynoceras sparsicosta* and *A. henleyi* show progressively ribbed inner whorls, and *A. lataecosta* is ribbed throughout

ethologist, has shown, fish of different species will feed on a coral reef in close proximity but without conflict. Likewise, the French biologist, Professor Grassé, looked for competition among the many species of French butterflies but found none.

Darwin does not seem to have distinguished very clearly between competition among the members of one species and competition between species. In general, specialisation prevents conflict here too. The one salient exception is that of predator and prey – Darwin's wolves, in fact. But even here it is not at all certain that selection occurs. Thus Canadian biologists studied the population of hares preyed on by lynxes and found that the numbers of each rose and fell together in a cycle lasting about nine years. The evolutionary effect was nil.

The postulate of competition makes more sense in the plant world, where vigorous species frequently drive out weaker ones, as every gardener knows. But even here there are phenomena which take some explaining. J. C. Willis back in the 'twenties, studied the Podostemaceae in Ceylon. They live in rapidly flowing water and he counted 160 different species, grouped in forty genera. Living in the same habitat, why did they not evolve to a common form? In fact there was an amazing variety, 'greater than any other family of plants'. (Water plants usually show little variation.)

5 *Proofs of Natural Selection*

The trouble with the theory of natural selection is that it is virtually impossible either to prove it or to disprove it. The philosopher Karl Popper insists that unless a theory is falsifiable (in the event of its being false) then it is useless and does not qualify to be called a scientific theory. He points to the Darwinian theory as a perfect instance.

We cannot go back in time to see for ourselves whether evolution always and everywhere occurred on the lines laid down by Darwin. However, since Darwin claimed that *all* changes arose by chance and all changes were selected, we can look for cases which do not fit the rule. It is the Darwinians who are really in the fix, since they can never prove that their rule has no exceptions. Actually Darwinians have made incredibly few attempts to prove their case in any scientifically convincing way, preferring to amass instances of natural selection at work. One of the few laboratory demonstrations was performed with the small mosquito fish, which appears in a pale grey and in a black form. Tanks were filled with equal numbers of both kinds of fish and penguins were allowed to prey on them. One tank was painted black inside, the other pale grey. As you will have guessed, the black fish survived better than the grey in the black-lined tank, the penguin taking 78 black fish to 217 grey fish. Conversely, more grey fish survived in the pale tank.

A natural experiment of a similar kind which is constantly cited, is the study of the peppered moth, *Biston betularia*, by H. B. Kettlewell. It resembles the lichen on trees so closely as to be almost invisible when resting on a lichened tree. About 1848, a black form materialised, named *carbonaria*, which was easily spotted by birds and hence eaten. The mutation kept recurring, however. But the smokes of the industrial revolution killed the grey lichens and blackened the trees, giving the black form an advantage and handicapping the grey form. By 1900 the black form, which had constituted only one per cent of the population in the mid-nineteenth century, comprised 99 per cent. Recently, owing to the Clean Air Acts, the drift has been reversed and in some areas the grey form in reasserting itself. This has been called 'the most striking example of evolutionary change actually witnessed'. And even 'Darwin's missing evidence'.

Subsequently more than seventy other varieties of moth were found to have darkened and the same is true of industrial areas in Europe, the US and Canada. But popular accounts of this phenomenon say nothing of the many reservations Kettlewell expressed. In particular, it is not clear why the disadvantageous form is not eliminated completely in a century, which is what Darwin would have expected. And why are the proportions surviving quite different in different species? Odder still, dark mutants appeared among ladybirds and spiders, even among ladybirds that are distasteful to birds in any case.

An experiment with plants that is often held up as being particularly convincing was launched in 1903 at the Illinois Agricultural Experiment Station and continued until 1927. A population of corn was grown, the parents for the next generation always being the plants with ears closest to the ground. At the start, the average height of ears above the ground ranged from 43 to 56 inches. At the end of twenty-four years, the average had become a mere eight inches, and this trait bred true. As a check, plants with the highest ears were also selected and bred; in this case the average height rose to 120 inches (10 feet!) by the end of the experiment.

Such experiments unquestionably show that selection can shift the genotype in respect of single characters – height, colour – controlled by a single gene. They tell us nothing about the complex changes involving numerous genes which comprise the main stuff of evolution, of which we shall encounter many examples in the pages which follow. Nor do they help us to understand how completely new organs such as the eye and the ear make their appearance. And of course they do not prove Darwin's three universals – that *all* changes are due to natural selection, are minute and arise by chance.

In all the thousands of fly-breeding experiments carried out all over the world for more than fifty years a distinct new species has never been seen to emerge. (I exclude one claim made in 1966 on the basis of a technicality.) Only in bacteria, where drug resistance emerges rather readily – too readily – and where the ability to subsist without some normal dietary component can emerge, do we see anything approaching evolution in the act.

In short, the case for Darwin's theory has never been conclusive. What has thrown it into serious doubt and sparked the launching of new hypotheses is the amazing advances in molecular biology during the past twenty-five years, and in particular the perfection of techniques for analysing genetic material at the molecular level. Evolutionists have been operating with a childishly primitive conception of genes and of gene mutations. It is turning out that genes are much more complex structures, with many more amazing tricks up their sleeves, than anyone supposed. In particular, they do not just sit there waiting for a mutation to occur, but are embedded in complex regulatory systems. Darwinism was not a bad first approximation but, just as Newton's laws had to be subsumed in Einstein's far more intricate theory, Darwinism is giving way to a far more sophisticated picture of how evolution occurs.

While Darwin's theory is shuddering under attack, the theory which has often been regarded as the rival one, namely the theory of Lamarck, is regaining a degree of credibility. It deserves a chapter to itself.

CHAPTER TWO

The Problem of the Giraffe

1 Of Ostriches and Warthogs

Since it first burst on the world Darwinism has been beset by a controversy which has become steadily sharper with the passage of time – the issue known as Lamarckism. In a nutshell, where Darwinism claims that all heritable variations are due to chance, 'Lamarckism' claims that changes evoked by the environment – such as the swelling of muscles on doing heavy work – can become inherited. This is known as the inheritance of acquired characteristics.

Actually this is not quite what the French naturalist the Chevalier de Lamarck suggested, but as the term 'Lamarckism' has become accepted I shall use it.

Darwin himself, as a matter of fact, was inclined to believe that such inheritance occurred and cited the reported case of a man who had lost his fingers and bred sons without fingers. He said that 'the tendency to vary is so strong that all the individuals of the same species have been similarly modified *without the aid of any form of selection*' (my italics). It was the neo-Darwinians, such as Haldane, Huxley, Fisher and Sewall Wright who, in the 1930s, made such an issue of the matter, backed by men like Simpson, Mayr, and Dobzhansky – all figures we shall meet later – and did so in terms so sweeping as to make a sceptic suspect they had an uneasy conscience. For instance, Professor C. D. Darlington of Oxford called it 'an evil theory' and said that to impugn Darwin's theory was 'ignorance and effrontery'. Harvard's Professor Bernard Davis is even more categoric. 'Except for those sceptics who are willing to discard rationality,' he says crushingly, 'Darwin's theory has now become Darwin's law.'

But let us look at an actual case: the classic example is the ostrich.

Very conveniently the ostrich is born with calluses on its rump, breast and pubis, just where it will later press upon the ground where it sits. These callosities are well defined in the unhatched chick. Curiously, it also has calluses on its ankles which are of no use to it, as it turns its foot sideways when sitting. They may, however, have been functional long ago when the ostrich had three toes and probably did not turn its feet over.

It is tempting to suppose that ostriches once developed these calluses in the course of life, much as the human hand becomes horny with toil, and that somehow the capacity to produce them became incorporated in the genome. This idea – the inheritance of acquired characteristics – has long been favoured by naturalists, who see many examples of similar processes. Thus the warthog goes down on its knees to root about with its tusks and has developed suitably placed hereditary calluses on its knees. Indeed the human foot is born with much-thickened skin as if in preparation for the wear it will experience after birth. Of course it is not just calluses but all pre-adaptations which are in question. How did wading birds acquire their long legs, or the elephant its trunk?

But while naturalists cling to the idea of inheriting acquired characteristics, which seems only common sense, geneticists reject it entirely, solely on the grounds that they know of no mechanism by which the genome could be appropriately altered by experience. Instead they ask us to believe that by pure chance, the genes needed to evoke the appropriate changes are created, or perhaps activated, and that this happens in quite a short space of time. In the case of the ostrich, they claim, by pure chance genes were developed which would cause just the required calluses to develop and no others, and that those birds which acquired such genes were so much advantaged that they survived preferentially. That takes a lot of believing even if you practise – like the White Queen – believing three impossible things before breakfast. I myself can only believe it on certain days and not a few biologists have failed to manage it altogether. Professor Waddington (whom we shall meet again in a moment) has emphasised that when circumstances modify an organism, the modification is of a kind which improves fitness far more frequently than would be expected on a random basis.

Some of the comments of the pundits are quite remarkably vapid. Thus all that the great Ernst Mayr, an ornithologist turned evolutionist, can find to say about the problem of the calluses of the ostrich is: 'It is not difficult to understand that such genes will accumulate in a population as soon as such a modification of the phenotype becomes of selective advantage.' Of course – but how do we know that it is of selective advantage or even where the genes come from? It is very nice to have a cushion to sit on but the human race does not die out where cushions are not available. Mayr's comment shows how little he has attempted to understand the problem.

As I said, the argument is not confined to calluses. For all the millions of structural and functional adaptations of living things, genes must have arisen to fill the bill, just when they were needed. One could perhaps imagine the emergence of a gene which thickened the skin generally – but one which thickens just those parts which will receive wear is too improbable. *A priori* it is more likely that something happened in the affected parts to start the ball rolling. Professor C. H. Waddington has remarked: 'It is doubtful whether even the most statistically minded geneticists are entirely satisfied that nothing more is involved than the sorting out of random mutations by the natural selective filter.'

This of course is the crux of the whole argument. It is easy enough to believe that selection favours certain modifications once they are established, and that it disfavours others. The whole question is: how do such precisely adapted modifications arise? The geneticists say: pure luck. Their opponents say: in response to a demand. The geneticists say: no one has ever demonstrated the inheritance of acquired characteristics unequivocally. But then no one has demonstrated that adaptive modifications arise *by chance*. They have been shown to arise, and to be favoured by selection, but how they arise is as much a mystery as ever. Brutal treatment of the genetic material will produce abnormal and defective forms. Prolonged selection will favour the development of small modifications. But the origin of complex modifications remains a mystery. For many biologists and most laymen it is not only easier but more rational to suppose that a mechanism exists of which we are ignorant than to suppose that evolution is a matter of chance. This is why the belief in Lamarckism has persisted so obstinately in the face of scorn and anathema.

In short, despite the attempts of the geneticists to treat the battle as won, actually the question remains wide open.

The great paleontologist H. F. Osborn summed the matter up as long ago as 1895 when he said: 'If acquired variations are transmitted there must be some unknown principle in heredity; if they are not, there must be some unknown factor in evolution.'

2 *Lamarck: Of Birds and Giraffes*

The doctrine of the inheritance of acquired characteristics has always been associated with the name of the French biologist the Chevalier de Lamarck.

Lamarck put forward his ideas on evolution in 1809 in his *Philosophie Zoologique*. Despite repeated attacks, they have been taken up again and again for nearly two centuries. The French, in particular, have never quite accepted that Darwin got it right.

Though Lamarck's name has become covered with contumely, he was in fact a great naturalist: his contributions to the classification of the invertebrates alone are sufficient to have earned him an honoured place in the history of biology. More than this, he can claim to be the first biologist to propose a theory of evolution, though of course he did not use the word 'evolution', which to eighteenth-century biologists meant the development of the embryo. Neither did Darwin, for that matter; he spoke of 'transformation'. Lamarck spoke of the 'march of nature.'

Actually the first person to suggest that the great 'ladder of nature' – an idea going back to Aristotle – implied an upward movement, a transformation of one species into another – was Erasmus Darwin, uncle of Charles. C. D. Darlington asserts: 'Erasmus Darwin originated almost every important idea that has since appeared in evolutionary theory,' including the inheritance of acquired characteristics. Unfortunately he buried them in his immensely long poem, *Zoonomia*. Lamarck, if not then the first to propose explicitly that 'these individuals originally belonging to one species become at length transformed into a new species distinct from the first', was certainly the first to devote an entire book to this idea.

At first Lamarck assumed that the sequence of forms was linear – that each creature had transformed into one slightly more complex and so on, indefinitely; but before long he realised that this could not be so and accepted the idea of branching evolution and the vanishing of many forms. (This was a bolder claim than it looks; the orthodox claimed that God could not have made mistakes.) No writer before Lamarck so well appreciated the importance of adaptation. He realised that, if there had only been one creation as the theologians averred, existing species must now be very ill adapted in view of the many geophysical changes which geologists were proving had occurred.

It was when he looked for a mechanism to explain adaptive change that he went wrong, as it is now thought, but not as wrong as is claimed. He did *not* say that environmental conditions affected heredity. He said: 'The environment affects the shape and organisation of animals [but] it does not work any direct modification whatever in their shape and organisation.' He pointed out that alteration of the environment would alter an animal's needs, and hence its *behaviour* and that this would lead to an alteration of its structure. And whereas eighteenth-century naturalists had thought that structure determined use, Lamarck realised that the reverse was true. 'It is the needs and uses of the parts which caused the development of the same parts, which have even given birth to them when they did not exist . . .'

These were brilliantly original speculations: unfortunately he did not support them, as Darwin did, by detailed experimental and observational work – and it was accumulation of facts which impressed the Victorian public, with its tradition of hard work.

In 1827 a giraffe was presented to the Museum of Natural History in Paris and became a source of wonder to everyone. Lamarck (then eighty-three) thought it a perfect example of how bodily construction was shaped by habit and environment. The giraffe, it was said, had acquired its long neck by repeatedly reaching up to crop the highest leaves. This is actually nonsense, from any point of view, for the neck of the female giraffe is two feet shorter, on average, than the neck of the male, so that the females would soon starve to death, bringing the genus rapidly to extinction. In point of fact, the giraffe needs a long neck because its legs are long: without it the animal could not reach the ground to drink or crop grass. (And if you have observed giraffes eating, you will have noticed that they spend more time grazing than they do tearing leaves off trees.)

To do Lamarck justice, he never made this assertion about the neck of the giraffe; it was the giraffe's legs which he discussed, and the number of biologists who have repeated the error of saying he wrote about its neck shows how seldom scientists go back to original sources. Actually, in his original writings he did not mention giraffes at all, but discussed wading birds.

It was Darwin, not Lamarck, who drew attention to the giraffe's neck. He said: 'The giraffe, by its lofty stature, much elongated neck, forelegs, head and tongue, has its whole frame beautifully adapted for browsing on the higher branches of trees. It can thus obtain food beyond the reach of the other Ungulata or hoofed animals inhabiting the same country and this must be a great advantage to it . . .' and so on.

Lamarck chose as an example the case of wading birds, with their long legs, and skin which webs the feet of ducks. 'Wishing to avoid immersing its body in the liquid', he said, the wading bird 'acquires the habit of elongating and stetching its legs.' The kangaroo likewise showed the effect of constant use in strengthening its back legs.

Lamarck also claimed that his theory explained the disappearance of organs with disuse, as in the case of the eyes of the mole or of cave-dwelling fish. For if using a structure develops it, not using it will have the opposite effect. Darwinians would say that the genes specifying the organ in question mutate or are lost and that since this does not cause a loss of fitness, no selective effect occurs. But this does not entirely fit the facts. Organs, such as eyes or legs, do not vanish suddenly or develop abnormal features, they slowly fade away. Take, for instance, the thigh bone of the whale, which has now shrunk to a relic about eighteen inches long, buried deep beneath its blubber. It is inconceivable that there was any evolutionary advantage to be gained once it had shrunk to the point at which it no longer disturbed the whale's streamlining. It seems rather that a regressive process has been set off which will doubtless result in its final disappearance, regardless of selection. Darwinian theory is unable to give a satisfying explanation of such regressions.

Likewise, the presence of even fully functional eyes could hardly handicap the cave-dwelling fish, or the mole, which does occasionally surface. The muscle which fills up the eye-socket actually takes more energy to maintain than the eye itself.

In Lamarck's day, French biology was dominated by Napoleon's protégé, the great Cuvier, who believed firmly in a single creation and postulated a number of catastrophes, of which the Flood was only the last, to account for the presence of extinct species in the rocks. Cuvier naturally poured scorn on Lamarck's ideas and after Lamarck's death, when he was called upon to deliver an eulogy, as President of the Academy of Sciences, he did not hesitate to misquote Lamarck in order to ridicule him. Lamarck was poor all his life and ended in a pauper's grave. We do not even know the names of his several wives and no one has written a definitive biography of him. He is one of the tragic figures of science.

Particularly unfair to him was Darwin, who skimmed through one of his books and pronounced it a farrago of nonsense. He had not, he said, gained a single idea from Lamarck. This was doubly ironical, for Darwin repeatedly toyed with the idea of the inheritance of acquired characteristics and, if it is so dreadful, it is Darwin who should be denigrated rather than Lamarck. (Darwin was equally grudging to his uncle, saying that his *Zoonomia* had merely 'anticipated the erroneous opinions of Lamarck'.) In the 1859 edition of his work Darwin refers to 'changes of external conditions' causing variation but subsequently these conditions are described as directing variation and cooperating with natural selection in directing it.

And in the famous passage about the giraffe, he actually refers to offspring which inherit the same bodily peculiarities as their parents or 'with a tendency to vary again in the same manner'. Every year he attributed more and more to the agency of use or disuse, and even wrote: 'The tendency to vary in the same manner is so strong that all the individuals of the same species have been similarly modified *without the aid of any form* of selection.' (My italics.)

By 1868 when he published *Varieties of Animals and Plants under Domestication* he gave a whole series of examples of supposed Lamarckian inheritance: such as a man losing part of his little finger and all his sons being born with deformed little fingers, and boys born with foreskins much reduced in length as a result of generations of circumcision.

In fact, following criticisms by Professor Fleeming Jenkin, who pointed out that new species could never arise by chance if inheritance was based on blending, he inserted a whole new chapter in the sixth edition of *The Origin*, reviving the idea. He would have done better to stick to his guns.

The idea that natural selection was the motive force in evolution was not original to Darwin either: Professor William Lawrence, FRS, had proposed it in 1822 before Darwin ever set sail. So when Darwin referred to 'my theory', as he frequently did in early editions, it is hard to know what theory he meant or could claim to have originated.

Professor C. D. Darlington, who is an out-and-out believer in natural selection, has nevertheless declared that Darwin 'was able to put across his ideas not so much because of his scientific integrity, but because of his opportunism, his equivocation and his lack of historical sense. Though his admirers will not like to believe it, he accomplished his revolution by personal weakness and strategic talent more than by scientific virtue.'

3 *Of Salamanders and Toads*

In the early part of this century a young Viennese investigator, Paul Kammerer, was conducting painstaking experiments with salamanders which seemed to some to show conclusively the existence of Lamarckian inheritance. This was ironical, since Kammerer was a Mendelian. Kammerer used two contrasting animals in his first series of experiments: a black Alpine salamander which bears two offspring at a time and does so on land, and a spotted salamander which produces from ten to fifty larval offspring at a time and these in water. Kammerer brought up spotted salamanders in an Alpine evironment and black ones in a lowland situation. They switched roles, the Alpine one producing numerous larvae and the spotted one two live offspring. What was more astonishing, this pattern persisted in subsequent generations.

For the next eleven years Kammerer continued his experiments, bringing up black and yellow salamanders on yellow and black soil. You can guess what happened: the black ones became yellow, when on yellow soil, the yellow ones black. And this too persisted. His findings having been confirmed by another worker, Kammerer received the prestigious Sommering prize. Could it be that Lamarck was right after all?

These were not the only startling observations bearing on inheritance made by Kammerer. He turned to the blind newt, *Proteus*. If *Proteus* is brought up in the light, it remains blind. But Kammerer tried raising them in red light, whereupon they developed eyes, showing that the hereditary information for creating eyes had not been lost, only suppressed. This was a major discovery, more wonderful at the time than it might seem today. When in 1923 Professor Ernest MacBride of Imperial College saw Kammerer's specimens at a special meeting of the Linnean Society in London, he called them 'the most wonderful specimens in my judgment which have ever been exhibited to a zoological meeting'.

But the work which was eventually to prove his undoing was carried out with the midwife toad, *Alytes*. Arthur Koestler has described the débâcle brilliantly in his *The Case of the Midwife Toad* and I cannot do better than follow his account.

Most frogs and toads copulate in water and the males have happily developed pads on their palms which enable them to grasp the slippery female more effectively. The male actually holds the female in his grasp until she spawns, which may take weeks. These 'nuptial pads' are dark in colour, roughened and equipped with tiny spikes. But the species known as *Alytes obstetricans* has chosen to mate on land and lacks these pads. 'Could *Alytes* have retained the power to produce pads?' was the question Kammerer asked himself. So he induced *Alytes* specimens to mate in water for several generations until they eventually developed – so he reported – nuptial pads.

It is important to stress that Kammerer himself did *not* regard this as an instance of the inheritance of an acquired characteristic but rather as an atavism, the recovery of a lost capacity. But William Bateson, the leading figure in British genetics in the 'twenties, did so regard it and the idea was one he could not tolerate. Bateson was not a lovable character. His son, questioned by Koestler, replied succinctly that many of father's colleagues 'hated his guts'. The fact was that Bateson had started out in life as a Lamarckian and had spent many months in the Karakorum desert vainly trying to find evidence to support the idea of Lamarckian inheritance. When he read Mendel's paper he was instantly converted to the Mendelian view of inheritance: it explained his failure. Like many converts, he was bitterly critical of those who held the view he had abandoned. The idea that, after all, there might be some shred of truth in Lamarckism was intolerable. He therefore determined to destroy Kammerer.

He had begun his campaign in 1910 by writing smarmily polite letters to Kammerer which led to his being invited to Vienna, where he was accommodated by the head of the institute in which Kammerer worked, Dr Hans Przibram, who treated him as an honoured guest, taking him to the opera and showing him many courtesies. But he did not see the famous nuptial pads, because the toads were not breeding at that time, though he saw the salamanders and other of Kammerer's experiments. 'He comes uncommonly near showing that an acquired adaptation is transmitted,' he wrote to his wife, adding 'Taking the whole series of experiments together I cannot really entertain the idea of fraud.'

Then had come the war: the staff of the Institute were all called up. Kammerer, excused military service because of a heart condition, worked in the censorship. His experimental animals all died and the Institute became a shambles.

In 1919, the war over, Kammerer published a detailed paper on his work with *Alytes*. Bateson at once wrote to *Nature* (in which a summary of the *Alytes* work had appeared) criticising the photographs of the pads and suggesting that Kammerer was publishing them in order to support Lamarckian inheritance. For good measure, he also hinted that Kammerer had faked his salamander results. When Kammerer sent him microscope slides of sections of the pads to examine, he suggested that the sections had been cut from normal pad-bearing toads. The head of the institute, Przibram, confirmed that he had personally been present when the sections were cut.

The issue intensified in 1923, when Kammerer was invited to lecture in Cambridge. He brought with him specimens of his work, including the only remaining *Alytes*, the rest having been lost in the war. This was closely examined by the Professor of Zoology at Cambridge, the Professor of Zoology of Imperial College, and numerous other distinguished scientists, including Professor Haldane. None of them suggested that the specimen was not genuine. Bateson, however, refused to attend the lecture. Kammerer impressed those present as a man of great integrity and charm. Bateson met him, however, in London and apologised for the rudeness of his attacks – only to renew his accusations after Kammerer had left.

The climax came in 1926, when Dr G. K. Noble, Curator of Reptiles at the American Museum of Natural History, visited the Vienna Institute and announced that the *Alytes* specimen was faked. (Gregory Bateson describes him as 'a ruffian'.) He had already criticised Kammerer's work and now he was irked by the sensational publicity which had attended the latter's lecture tours in America.

Noble showed that the frog's claw had definitely been injected with Indian ink, and Przibram was forced to agree that this was so. Six weeks later, Kammerer committed suicide, his great book on the transformation of species uncompleted.

The fake was such a crude one that Kammerer could not conceivably have carried it out himself. Was it done by some well-meaning assistant, just before Noble's visit, because the markings on the fifteen-year-old specimen were fading? Przibram mentions a 'madly jealous colleague' who later had to be locked up. Could he have done it? Was the motive political? Since Kammerer's work was so well thought of in Russia, could a Nazi have wished to discredit it?

The fact that the Russians did not withdraw the offer they had made to Kammerer to set up a special institute for him suggests that they were convinced of his personal integrity. There was, after all, little need for a man who had several major discoveries to his credit to resort to fraud.

Discredited on one matter, the whole of Kammerer's work – on salamanders, on *Proteus*, on lizards and on sea-squirts – was discredited too. No one has ever thought it worth attempting to repeat it. The extreme difficulty of breeding these creatures in captivity and the many years which must pass before enough generations have been bred to prove inheritance, were enough to deter scientists even before the débâcle. But it is inconceivable that all Kammerer's work was faked, and the results still call for explanation. And even if it was faked, this does not actually disprove that Lamarckian inheritance can occur, it simply leaves the matter open.

But soon another blow was to be dealt to Lamarckian doctrine, a left-handed one.

4 *Of Wheat and Cows*

In Russia, Lamarckism was de rigueur. It was central to the Marxist view that people and society were malleable: society could be made into anything you wanted; people too. The existence of hereditary factors, which might prevent such change, was therefore hotly denied. Mendelian theory, therefore, was considered 'bourgeois deviationism'. 'The theory of heredity contributes nothing to the evolutionary process,' declared Smirnov in his influential *Essays on the Theory of Evolution*, published in 1924.

Kammerer's work had therefore aroused great interest in Russia, and for three years before his death he spent much time travelling in Russia and lecturing. The Russians decided to set up a research institute for him, and did not withdraw their offer after the exposure of the fraud. It was Kammerer himself who wrote saying he no longer felt he could take the job on.

This situation was exploited by an extraordinary figure, a Russian peasant named Trofim Lysenko, who succeeded for some thirty years in discrediting Mendelian genetics in Russia. It is a story worth telling in some detail if only for the awful warning it constitutes.

Lysenko was not an attractive character. A journalist who visited him, after the publication of his first paper, in 1927, wrote: 'If one is to judge a man by a first impression, Lysenko gives one the feeling of a toothache. God give him health – he has a dejected mien. Stingy of words and insignificant of face is he; all one remembers is his sullen look creeping along the earth as if, at the very least, he were ready to do someone in.' And do someone in he eventually did – N. I. Vavilov, the brilliant leader of Soviet genetics, President of the Plant Research Institute, founder of the great Russian plant collections, a man of tremendous energy and wide interests.

Lysenko started in a small way in 1928 with a paper on the 'vernalisation' of wheat. Actually, there was nothing new in this. Winter varieties of wheat can be planted in the spring if they are allowed to germinate slightly in the fall but kept from developing by holding them at a low temperature. This had been discovered by an American, J. H. Klippart, as far back as 1857 and forgotten. Lysenko claimed it as a sensational discovery, making it possible to exploit the colder areas of Russia for wheat growing. Subsequently experiment showed that vernalisation did not raise yields and it was quietly dropped.

Lysenko developed, however, a new investigative technique. Instead of proving his claims by his own experiments, he sent questionnaires to fifty collective and state farms asking if vernalisation had improved yields. The directors all dutifully replied that it had. Of course, the gross amount of wheat turned in remained unchanged, but nobody noticed that. Lysenko was to use this kind of 'proof' again and again.

He also claimed to have developed new varieties of wheat resistant to disease, to 'lodging' and so on – but they did not stand up to tests. He maintained that crosses should be made within a single variety and not with other varieties, as was normal. Then he announced that plants should not be self-fertilised. An army of collective farmers was to remove the anthers from every spike with tweezers, so that they would be fertilised by wind-borne pollen. As his fame mounted he was allowed to determine agricultural policy. He advocated the planting of southern varieties of wheat in Siberia on unploughed land. It was a disaster. He organised the planting of beets over enormous areas of parched land. They all died. Later editions of his works omitted all reference to summer beets. His 'branched wheat' was advertised as revolutionary, but turned out to be poorer than ordinary varieties, both in yield and disease resistance. He produced wild ideas about fertilisers. One of his co-workers even abolished hormones, which were declared to be 'Mendelism-Morganism'. Finally, Lysenko announced that he had solved the problem of speciation.

The culmination of Lysenko's amazing list of pseudo-discoveries was his claim to increase the butter-fat content of milk by crossing Jersey cattle with local breeds. Lysenko was convinced that the fertilised egg (the zygote) develops not in accordance with Mendelian principles but in the way the egg-cell thinks most profitable to itself. So, if a large bull with genes for high butter-fat is crossed with a small cow, butter-fat content in the offspring will nevertheless be low, because the egg feels that a large calf would have difficulty in emerging from a small cow. But if a small bull is crossed with a large cow, the egg feels no such apprehension! He advocated that all crosses therefore be made in this manner.

How did this half-educated man succeed in dominating Soviet agrobiology for more than thirty years? Monuments were erected to him, busts of him were sold, hymns were even written to him. He was dubbed 'Professor' although he had never taught. He was made a Hero of the Soviet Union.

His strategy was to politicise the issue. Those who opposed his plans on the farms themselves were 'kulak wreckers' and 'class enemies'. Orthodox geneticists were Trotskyites, bandits, agents of capitalism, double-dealers and even 'maidservants of Goebbels'. How they could be both fascist and Trotskyite at the same time was not explained. The language used in these attacks is so intemperate that one wonders how people could listen without writing the speaker off as half demented. For instance, Lysenko's alter ego, I. I. Prezent, objecting to the publication of a critical article in the journal of the Academy, described the editor, Uranovsky, as 'crawling on bent knees after the latest reactionary word of "scientists" abroad', and saying that he 'was able, as his last malicious spit at our Soviet science and, so to speak, in the manner of servile grovelling on a world scale' to publish the article in question. After attacking 'another Trotskyite bandit', the geneticist I. I. Agol, he turned to the human geneticist S. G. Levit 'who provided man-haters with "material" on the alleged "hereditary foredoom" of man, [and thus] earned a kiss from the hardened opponents of Marxism in science'.

Stalin cried 'Bravo, comrade Lysenko', after Lysenko's first major speech at an agricultural congress and thereafter Lysenko could do no wrong.

The next phase opened in 1937 with a series of attacks of increasing virulence on Vavilov, culminating with his arrest in 1940, followed by the arrest of many colleagues, several of whom died in prison. Vavilov's trial took place in 1941: at it he was accused not only of 'sabotage in agriculture' but of belonging to a rightist conspiracy and spying for Britain!

After he had spent nearly a year of solitary confinement in an underground death cell, his sentence was commuted to ten years' imprisonment. He was transferred to a prison in Saratov; there he died of malnutrition in 1943. Twelve years later he was 'rehabilitated'.

The stream of discoveries in genetics, and in particular the establishment of the structure of the DNA molecule in 1953 made the anti-Mendelian position harder and harder to maintain. Lysenko's patron Stalin was dead and a new generation of biologists had grown up. Khruschev too had been a supporter of Lysenko, but following his demotion in 1964 more and more people felt enabled to speak out. Articles attacking Lysenko's position began to appear. The hollowness of his claims was gradually exposed by the facts. In 1965 Lysenko was dismissed from his post as director of the Institute of Genetics.

5 *Dead But Won't Lie Down*

Despite this unhappy history, a minority of biologists have persisted in carrying out experiments to test the possibility of Lamarckian inheritance. And if the number of such experiments is small, at least it is not as small as the number of experiments showing the chance emergence of favourable mutations above the one-gene level. It is a striking, but not much mentioned fact that, though geneticists have been breeding fruit-flies for sixty years or more in labs all round the world – flies which produce a new generation every eleven days – they have never yet seen the emergence of a new species or even a new enzyme.

Fifty years ago, for instance, one Harry Schroeder conducted an intriguing experiment with the willow-moth caterpillar. This caterpillar places itself on a leaf and rolls the leaf round itself before pupating, fastening it down with a web. Normally, it starts by drawing the tip of the leaf over itself, but Schroeder, with fiendish cunning, systematically cut off the tips of all the leaves on which caterpillars had taken up position. Sensibly enough, they responded by drawing the side of the leaf over instead. When these caterpillars had produced another generation, Schroeder found that, of nineteen offspring, four drew the side of the leaf over, not the tip, when their time to pupate came around. It may be said that this was inheritance of an acquired behaviour, not a structure, but there may not be much difference from a genetic point of view, as we shall later see.

Perhaps the same might be said of an unique series of experiments by Frederick Griffiths, who placed rats on slowly revolving turntables for periods of up to one and a half years. When the wretched animals were freed their heads constantly flicked in the direction in which they had been rotated, and their eyes flicked also. This flicking automatism reappeared in their progeny.

Some of the most convincing experiments have been done with plants. For instance in 1962 Alan Durrant at the University College of Wales, Aberystwyth, induced changes in the flax plant by cultivating it with different types of fertiliser. Some plants became heavier and larger, others lighter and smaller. Astonishingly, these trends persisted for several generations. A few years later, J. Hill, at the Welsh Plant Breeding station, got rather similar results with tobacco plants, the flowering time also being affected. Durrant's work was carried forward by Chris Cullis and these lines of plants are still being propagated as I write, almost twenty years later. Some of the plants have produced offspring of the original form, but these in turn have given rise to unusually large and small plants. More significantly, Cullis has shown that when the large and small plants are crossed, the offspring follow the standard Mendelian pattern of inheritance (that is, large, medium and small plants appear in the proportions of 1:2:1). This makes it certain that a genetic change has been induced.

Some sceptics say that plants may carry on in this unseemly manner but that does not prove that animals do. In the field of animal experiments, some of the most convincing experiments have been done in the field of immunology. One of the first was that of Guyer and Smith who, between 1918 and 1923, performed experiments in which they ground up the eye-lenses of rabbits and injected it into fowls. When the resulting liquor was injected into rabbits, their offspring were born with small or defective eyes or with none at all, and such defects reappeared during nine succeeding generations, growing worse rather than better. Eye-lens material was chosen because it is well known to provoke immune reactions. However, some would say that immune reactions constitute a special case, since the immune system is specifically designed to recognise external influences.

Perhaps the most convincing, or anyway puzzling, experiments were carried out by Conrad Waddington, of Edinburgh University, about 1940. He exposed fruit-flies to heat-shock and produced some mutant flies which lacked the usual cross-veins in their wings. When he heat-shocked the next generation this mutation appeared more frequently, and as he continued with subsequent generations finally cross-veinless flies were appearing even when no heat-shock was administered. This looked so uncommonly like Lamarckian inheritance that Waddington, as a highly orthodox biologist, was disturbed. Since he was imaginative as well as orthodox, he formulated a theory, which he called genetic assimilation, to explain the results.

He started from the observation that, in biology, many processes seem independent of the nature or strength of the process which triggers them, and once triggered, they proceed to their destined end unaffected by minor influences around them. For instance, in the developing embryo, cells which have been triggered to become skin cells do so. You do not get cells which are halfway between skin cells and muscle cells, or any other type. It is as if the impulse, once launched, flows along a channel to a predestined end. He compared this with the stability of species, which remain very close to the master pattern despite being exposed to different environmental influences. Even if you induce mutation by some violent process, the mutants often revert to the original pattern.

Thus he saw the calluses on the ostrich as the result of such a process, appearing in much the same form whether the ostrich has lain upon rough or smooth surfaces. The production of the calluses, he points out, must be under genetic control. So we only have to assume that the genome at some point takes over the triggering of the channelling process.

This is, in reality, a quasi-Lamarckian explanation, though Waddington was careful to deny being a Lamarckian. What he did was to propose a mechanism by which acquired characters could become fixed in inheritance.

Ironically enough it later turned out that Waddington's experiment was not as Lamarckian as he thought; the heat treatment had caused the production of some unusual proteins, and transmission of the effect ceased after a few generations as these became diluted or died away. A later experiment, in which ether, instead of heat, provided the shock, went much the same way.

It was not simply that no one had demonstrated a mechanism: it had become a rigid dogma that information could not enter the genetic material. Weismann had started it by declaring that the germ-plasm was unalterably separated from the body cells. This turned out to be quite untrue, as can be seen from experiments in which the lower half of worms is cut off, including the sexual organs, whereupon they regenerate a new lower half, complete with sexual organs, from body cells. The dogma was restated by Francis Crick, who with Jim Watson, deduced the structure of the DNA molecule and the code by means of which it transmits genetic information. He declared that information could travel from DNA to protein but not from protein to DNA. So it was one of the biggest bombshells in the history of biology when the youthful Howard Temin at Wisconsin showed that viruses could carry genetic material into host cells and embed it in the host DNA, where in due course it would give rise to more virus, using the host-cell machinery for the purpose of making and assembling the raw materials. They manufactured a special enzyme for the purpose; Temin called it reverse transcriptase. Hotly disputed at first, Temin's experiments were soon confirmed, and in the remarkably short time (by the standards of the Swedish Academy of Sciences) of five years he was the recipient of a Nobel prize.

With the dogma thus modified, the philosophical objection to Lamarckism was removed. What continued to be lacking was a mechanism by which the phenotypic change, if any, could generate the information and transmit it to the genetic material. It was not simply that no mechanism could be demonstrated, no one could even imagine a way in which it could be done.

Then in 1979 the issue made the headlines with the publication of a book by the young Australian biologist, Ted Steele, proposing a mechanism and with the publication in the following year of the results of an experiment he conducted with Canada's Reg Gorczynski, in which a line of mice seemed to have acquired, and transmitted to their offspring, a certain immunological tolerance. Steele got the idea that the subject might be worth re-investigating from Arthur Koestler's book *Janus: A Summing Up*. 'I had a feeling I couldn't articulate about traditional evolution,' he remembers. 'It just didn't smell right. Koestler crystallised the problem for me.' He was twenty-eight at the time – it was June 1978 – and the following year he published a short book entitled *Somatic Selection and Adaptive Evolution*. Starting from Howard Temin's 1971 discovery that viruses can carry genetic material into the genome, Steele proposed that mutations can occur in body cells, be spread by viruses to other body cells and thus multiply, and finally be transmitted by viruses to the germ-cells, the egg and sperm, thus being passed to later generations.

It was Steele's friend and colleague Reg Gorczynski who first thought up an experiment by which Steele's ideas could be tested and who provided the laboratory to do it in. He chose the field of immune tolerance. As you will know, the body of men and animals tends to reject grafted material. The cells are recognised as 'foreign' by the immune system and are destroyed. But in the 'fifties Sir Peter Medawar demonstrated that if foreign cells are injected into a newly born mouse, later in life it will accept grafts from the same source. The injected cells have been catalogued, incorrectly, as 'self'. Medawar received the Nobel prize for this work, and the picture which he took of a white mouse with a black patch, grafted from another mouse, has become part of the iconography of biology.

What Gorczynski proposed to show was whether this tolerance could become inherited. Using exactly the same design as Medawar, they found that 50 per cent of the offspring of the tolerant (male) mice were likewise tolerant in the next generation, and the grandchildren were tolerant in from 20 to 40 per cent of cases.

The mechanism by which the body acquires such tolerance, at the genetic level, is unknown. Is there a gene for tolerance? Orthodox theory says just the reverse: the genes regulate immunity not tolerance.

Sir Peter Medawar invited Steele to pursue his experiments at the MRC's Clinical Research Centre, observing 'Something funny seems to be going on.' And he added that, if confirmed, Steele's findings would be 'a landmark in the history of biology'. However, a team of scientists including Medawar and Professor Leslie Brent of St Mary's Hospital Medical School have recently exactly repeated Steele's original experiments and concluded that no evidence was forthcoming for the inheritance of tolerance.

But I suspect the controversy will continue.

Although for more than a century Darwinism and Lamarckism have been presented as rival theories in reality they are not wholly incompatible. If a variation should occur for Lamarckian reasons it will be favoured by natural selection. The theories differ essentially on the question of whether the variations *only* occur by chance or whether they also occur in response to a need. Philosophically, the second alternative seems more likely for it is more efficient. There must have been continuous selection pressure for any mechanism which would facilitate the supplying of needs. Cybernetic theory, too, teaches us to look for a feed-back from environment to the organisms in that environment. The absence of a known mechanism is insufficient reason for rejecting the possibility.

In discussing the origins of the eye, Professor Wolsky in his book *The Mechanism of Evolution* comments that photosensitive organs seem to have appeared at those points on the body where the light fell most intensely, adding that this seems to imply Lamarckism. In a similar manner, the calluses of the ostrich appeared where the pressure was greatest: the skin did not thicken all over. The same is true for almost every adaptation: it appears not only when it is needed but where it is needed. The laws of chance would lead us to expect many more misses than hits. Evolutionists reply that the misses would not be selected for, but this is not altogether true: an eye on the side of the head would be of *some* survival value even if it were not so immediately useful as an eye on top of the head. While chance *could* perhaps explain such modifications, response to need explains them much better. In the absence of decisive evidence, the probabilities point to Lamarck rather than Darwin.

One can visualise that, when an ostrich lies on rough ground, or when the human foot is repeatedly stressed by contact with the ground, changes occur in the skin cells which affect the cell nucleus and cause it to bring about multiplication of skin cells and cornification or hardening. Though one can visualise it, the mechanism by which this occurs is not known. Nor do we know, in any detail, why muscles develop when frequently used, or why sweat glands become more numerous when you move to a hot climate. Still, it must be a genetic mechanism and so it is not inherently unlikely that changes in it might be transferred from body cells (in this case those of the skin) to the germ-plasm.

It is much harder to see why the legs of wading birds should grow longer, or flaps of skin develop between the toes of ducks and the fingers of bats. It is still harder to see what kind of stimulus could evoke the evolution of a complex organ such as the eye. A comprehensive Lamarckian theory would also have to account for physiological innovations, such as the provision of the appropriate pigments for colour vision. One almost sympathises with Lamarck's notion that it was the desire of the bird to keep its body dry, of the bat to fly and of the reptile to see in colour that triggered the process. But then one remembers the complex reactions which led to the formation of blood: one can hardly imagine the fishes wishing for more efficient blood! The trouble is, no one has tried to classify the possible kinds of mutation in a philosophical sense. We cannot be sure that they will all yield to the same kind of explanation.

If a mechanism by which external influences could affect the genetic material in body cells could be discovered, the views of Darwin, Mendel and Lamarck could probably be merged in a new synthesis.

However, there is another sense in which Darwinism and Lamarckism are in violent opposition. For Darwin believed, in the last resort, that the environment determined the way its occupants would evolve, whereas Lamarck, who believed that the creatures chose their environment, left the course of evolution open to free will and individual choice. These two views thus foreshadowed the political controversy which was about to flare up between Marxism, which claimed man could change his social environment and hence himself, and the Christians who held that God had preordained each man's role in life. And that, in a sense, is the battle between Darwin and Lamarck in human terms. It is fought not on a basis of fact but on a basis of conviction.

Marxists clove to Lamarckism because they profoundly believed that man could shape his own destiny. But there is a deep irony here, for Lamarck is the one who gives scope to *individual* choice, which is what humanists say they believe in. The philosophers do not seem to have got their arguments quite worked out.

CHAPTER THREE

Patterns of Living

1 From Water to Land

In October 1980, as I related in my Introduction, a conference was held in Chicago on one of the hottest issues in evolutionary studies. The respected magazine *Science*, organ of the American Association for the Advancement of Science, called it 'an historic conference' which 'challenges the four-decade long dominance of the Modern Synthesis'. 'We all went home with our heads spinning,' said one participant. 'Clashes of personality and academic sniping created palpable tension in an atmosphere that was fraught with genuine intellectual ferment,' *Science* reported.

The issue under debate was the jerkiness of the evolutionary story. Is the jerkiness merely an illusion created by the spotty nature of the fossil record, or is it the consequence of the jerky mode of evolutionary change? But beneath this lay a deeper issue. When we survey the evolutionary story, from the first multicellular creatures up to man, we soon get the feeling that from time to time there was a dramatic change of plan – and indeed of lifestyle – that is quite inconsistent with the slow accumulation of imperceptible changes upon which Darwin based his theory. Evolutionists call such discontinuities 'saltations' (that is, jumps).

The study of evolution in this large sense is sometimes called mega-evolution, in contrast with the micro-evolution of species; the term macro-evolution is also used but is ambiguous, since some workers use it for evolution only slightly above the species level. However this was the word used by the conference. Old-guard biologists deny that major saltations occur, so in this chapter let us look at some of them. What is the evidence? How gradual or sudden were they?

The most obvious and striking of these major steps was the step from sea to land, a step taken some 360 million years ago. Suddenly, four-legged air-breathing creatures appeared – quite unlike the scaly, limbless, water-breathing fishes which had been the most prolific creatures up to this time. The move from sea to land only became possible when a land-based food supply became available. In the Silurian and Devonian periods the land was a lifeless desert, except perhaps on the shores of some freshwater basins. Then the first plants, the psilophytes – now long extinct – appeared: small objects half an inch in diameter or less with rhizomes and branched shoots on which leaves gradually evolved – not wholly unlike the mosses we know now. By the Lower Devonian they were flourishing, to be followed in the Upper Devonian by the club mosses and by the horsetails, some of which grew to the size of modern trees.

We can visualise annelid worms living in the wind-driven detritus at the edge of lake shores, and the first spider-like insects and millipedes appearing.

The climate was variable, with rather wide daily variations too, so that pools and lakes would dry up, encouraging the fish in them to gulp air. Some of them discovered how to force air down their throats by pressing the palate against the floor of the mouth. They probably also absorbed air through their skins, as do the modern gobies, the skins of which are richly supplied with capillaries, as are the arches between the gill slits and the interior of the mouth itself. We know that one genus, *Boleophthalmus*, developed 'respiratory papillae' – nipple-like projections on the skin capable of absorbing oxygen. (This prompts the thought that if Fate or Evolution had favoured this device, perhaps we too should now be breathing through our skins.)

In short, it was not the need to breathe air which impeded fish from invading the land. The modern lung-fish is quite capable of living for months in the mud in a dormant state, breathing only air. And it can even walk on its fins! But it shows no signs of permanently abandoning its watery habitat. So why did the fishes invade the land? No one knows.

The real obstacles to such a move were the massive structural changes needed to make life on land worthwhile. To begin with, the fish would need legs simply in order to relieve the pressure of its body on the ground, which would compress the lungs.

Equally importantly, the land animal needs a strong pelvic girdle. The fins of fishes are attached only to bony plates beneath the skin and could not support the weight of the body until a link had been provided to transmit their support to the spine. There were problems with the front suspension too, for in fishes the forward fins are firmly linked to the skull. Turned into legs, the animal would have to move its head from side to side with each step, so a new system of suspension had to be provided. Finally, since the weight of the body was no longer taken by the water, the spine itself needed strengthening.

We are all so used to the idea of bone that it is hard for us to realise what a milestone the creation of bone was. Without bone, or something very like it, many terrestrial creatures could not support themselves against the drag of gravity. The American biologist T. H. Frazzetta recalls how the mystery of bones was what first turned his attention to the subject of evolution. 'I was drawn to them for the very unscientific reason that their variety of ornate surfaces, flowing twists and turns, odd-shaped apertures, hooks and bumps, made them, in my eyes, exquisitely fashioned works of sculpture,' he writes.

Bone has a precise, indeed an unique, structure, being composed of mineral and living matter interspersed. The strength of bone comes from the mineral component: crystals of hydroxyapatite; the adaptability from the living collagen. The two are arranged in specific patterns, with spaces reserved for living bone-making cells and for blood vessels.

You may be one of those who think of bone as inert, stony, almost eternal. In fact it is highly mobile, almost fluid on the evolutionary scale. Bone-building cells add to it here, bone-destroying cells erode it there, until it is sculptured into a different form, even in the span of a single lifetime. On the evolutionary scale, of course, much bigger changes are possible. If one of the larger bones is sliced in half, it is seen to contain a spongework of criss-crossing sheets, the trabeculae, which align themselves in precisely the best way to absorb the stresses to which that particular bone, in that individual, is being subjected. Like the network of girders which support a bridge or a structure like the Eiffel Tower, this gives strength for a minimum of weight. In addition the major bones contain a cavity, lined with a special sheath, which generates the blood cells needed by the blood. Human blood cells have a life of only 120 days, and you and I rely on our marrow providing a stream of replacements. Another membrane covers the exterior. (It is from the internal and external sheaths that the bone-making cells come.) Then there is the mystery of joints, with their capsule of cartilage and their remarkable lubricant, the synovial fluid.

It is obvious that the creation of bone required not one but a whole burst of mutations, all integrated to a single end – an incredible thing to happen by chance even if nothing else had been going on.

It is odd, too, that bone did not start in the spine, as you might

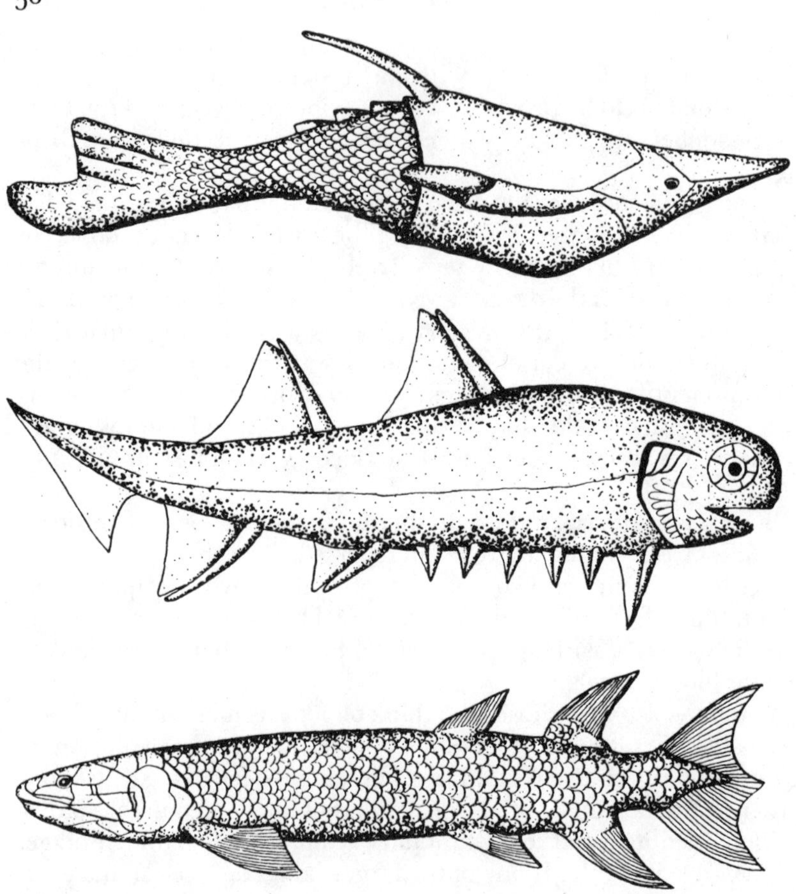

No evidence has been found of intermediate forms between fishes and amphibians. From the left, top to bottom, six specimens from the Devonian: *Pteraspis* and *Climatius*; the crossopterygian *Eusthenopteron*, a relative of the land vertebrates; *Coccosteus*; *Phamphodopsis*; *Bothriolepis*, a creature with a jointed exoskeleton including 'arms'; and *Ichthyostega*, the first amphibian of which anything is known

imagine, but on the surface of the head, like a kind of armour – so-called dermal bone. How it then took over the spine is a mystery. We have one clue. Dermal bone is formed within a mass of connective tissue; but the deeper skull bones, the ribs, limbs and part of the vertebrae are first sketched out in cartilage. As the embryo develops, the cartilage is gradually dissolved away to be replaced by bone. In other words, the creation of 'endochondral bone' called for still further coordinated genetic innovation. Evolution certainly likes doing things the hard way. In this respect if no other the vertebrates achieved a 'saltation'. The origin of scales (from which teeth and feathers were to derive) is another mystery story.

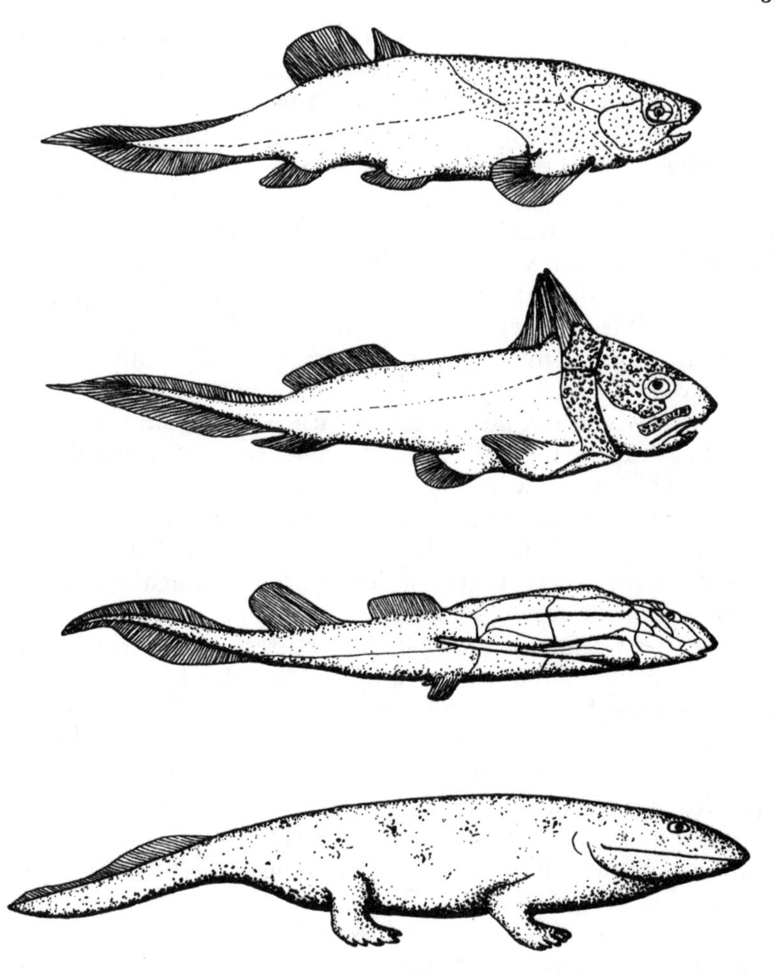

And so on to the next step, because land animals must also protect their body from drying out, by swapping scales for an impervious skin. Actually, the skin of some modern amphibians is quite sophisticated: it admits water when the creature returns to that element, the increased permeability being under hormonal control. We do not know if anything of the kind occurred in primitive amphibians. Land animals also need to protect their eyes from drying by a flow of tears and need an eyelid to protect it from dust particles. Similarly the nose must be protected by a supply of mucus.

The land animal must also change its sense organs. It no longer needs the curious organ which runs along its side called a lateral line, and this is converted, by an amazing series of steps which I shall shortly describe, into the ear. The eye, too, changes, since the refractive index of air is different from that of water and no doubt there are modifications in the sense of smell, though I doubt if anyone has studied that.

And then, of course, there is the problem of the legs themselves. Before ever the fish reached the land the structure of its fins began to change. Instead of rays, a series of bones corresponding to the tibia, radius and ulna of the arm appeared. Digits, tarsals and metatarsals evolved (so it is now generally conceded) as wholly new structures, though the point – unwelcome to Darwinians – was hotly contested in the 1930s.

The fish which decided to remain fish very sensibly converted their lungs into swim-bladders with which they could regulate the depth at which they swam.

Though we have this clue in the bone-structure of the crossopterygian fin there are no intermediate forms between finned and limbed creatures in the fossil collections of the world. Once again the critical evidence for gradual evolution is missing.

The earliest definitely four-footed creatures known were found in strata some 370 million years old in Greenland, which at that time was not icy but had a mild climate. Known as Ichthyostegids, they possessed a five-toed foot but retained the fishy tail and the lateral line of their fishy ancestors. Their skulls, however, were already typically amphibian and their jaws were equipped with teeth. About three feet long, they probably lived in shallow waters and impaled small fish on their sharp teeth. They are the nearest we can get to a 'missing link' in this context. From them derived, eventually, not only the reptiles but the salamanders and frogs.

By late Carboniferous times many other amphibia were flourishing; their vertebrae were typically amphibian but they were often legless or had only weak legs. Whether they came from a different stock from the Ichthyostegids no one knows. Most paleontologists doubt if the five-toed foot would have evolved more than once, to say nothing of the vertebrate type of skeleton. But the

Skeleton of *Ichthyostega*

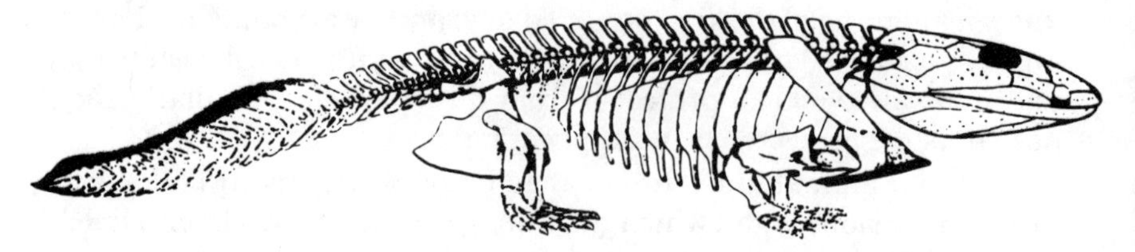

Swedish expert E. Jarvik, the leading authority, firmly maintains the contrary. Minute structural details are cited in support of each view. It is the sort of insoluble controversy on which paleontologists thrive.

All that need concern us is the larger question of whether such an impressive array of coordinated changes could have taken place by chance, and have done so without leaving in the fossil record a single intermediate form to prove the point. As Darwin complained: 'Where are the infinitely numerous transitional links' that would illustrate the action of natural selection? Not here, at any rate.

Among the odder features of amphibian development is the fact that the *Stereospondyli*, a group which developed in the Triassic after something like 150 million years of amphibian existence, returned to the sea and lived an almost wholly aquatic existence. It was a pattern which would be followed later by some reptiles, such as the plesiosaurs, and subsequently by some mammals, such as whales and dolphins. And in these last two cases, the change of direction is even more surprising, since reptiles and mammals were fully adapted to a life on land, whereas the amphibians were playing it both ways. Moreover those labyrinthodonts which persisted into the Permian also returned to the water.

Two forms of plesiosaur: *Kronosaurus* (top) and *Elasmosaurus*

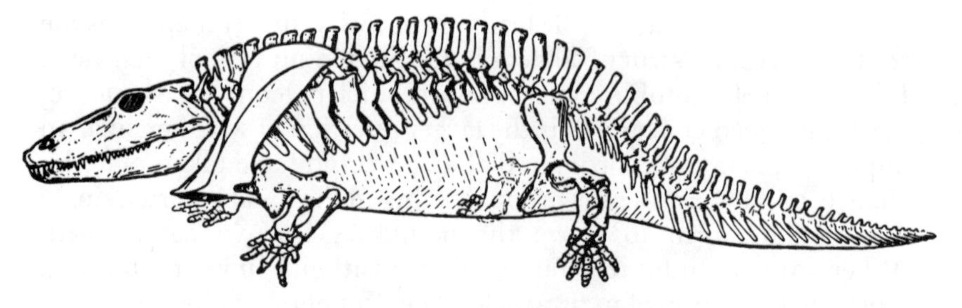

Skeleton of the Permian Labyrinthodont amphibian *Eryops*

The only conclusion one can draw from this is that amphibians lived on land because they chose to and not because they had to. And the same applies even more strongly to the reptiles and mammals just mentioned. The first fish could have used their walking fins to stagger back to the sea, or, better still, have learned to avoid being stranded in pools. It would surely not have been difficult to avoid the surf which warns you the shore is near. Thus even if we accept that natural selection generated the legs and lungs, we have not explained the change of lifestyle. The only reason there was selective value in developing such organs was because the fishes were attempting to live on land instead of staying comfortably in the sea.

Just as obscure as the origins of the amphibians is the branching of the frogs and salamanders (Anurans) from the amphibians. The fossil record is scanty until the Cenozoic, by which time the forms were quite modern. The most we can say is that they came from the main group of amphibians, the Labyrinthodonts (so called for their strangely fissured teeth) as did also the reptiles. Some preliminary moves towards the typically froggy structure, notably in the flattened shape of the skull, occurred in the late Carboniferous. Then, at the start of the Jurassic, there was a sudden and complete change to the modern type of Anuran, a change so successful that they have remained successful for 200 million years and have spread far and wide over the earth. In the late Carboniferous many other amphibia of obscure origins flourished, such as the Lepospondyls, with good backbones and sharp teeth, but weak legs or none at all. They have been a subject of controversy for thirty years.

Here, if this were a textbook, I should have to tell you how the amphibians gave rise to the reptiles, but I shall confine my account to a single feature of reptilian innovation, the egg – or rather, since fishes lay eggs of a sort, the amniotic egg – without which of course we cannot begin to understand how the birds contrived to emerge. It is one of the wonders of evolution.

2 Private Pools

The egg which graces our breakfast table is a minor miracle. To begin with, it looks so beautiful, both in shape and hue: beige or white in the case of hens' eggs, but blue or green and spotted with brown in the case of many other birds' eggs. Besides being smooth it is rigid enough to protect its cargo while not being so hard that the chick will be unable to peck its way out. The shell also is pervious to gases, so that the chick can breathe – which is one reason why one should not commit the folly of putting eggs in the refrigerator. Who wants eggs that taste of cheese, or worse?

Suspended in the middle of the egg is the yolk, supported by threads. You can rotate the shell of the egg twenty times without disturbing the yolk: the threads just wind up. The medium in which the yolk floats, the white or albumen, is remarkable too. As every cook knows, it will turn from a transparent liquid to a white solid on heating, and so serves to stiffen many edible mixtures.

I am speaking of course of the bird's egg as it exists today. The reptilian egg, as it first emerged, was slightly different. It contained the large yolk which served to nourish the developing embryo. It also contained two sacs, the amnion, filled with liquid and containing the embryo, and the allantois, which receives the waste products produced by the embryo while it is in the egg. It was however very

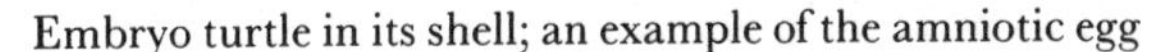
Embryo turtle in its shell; an example of the amniotic egg

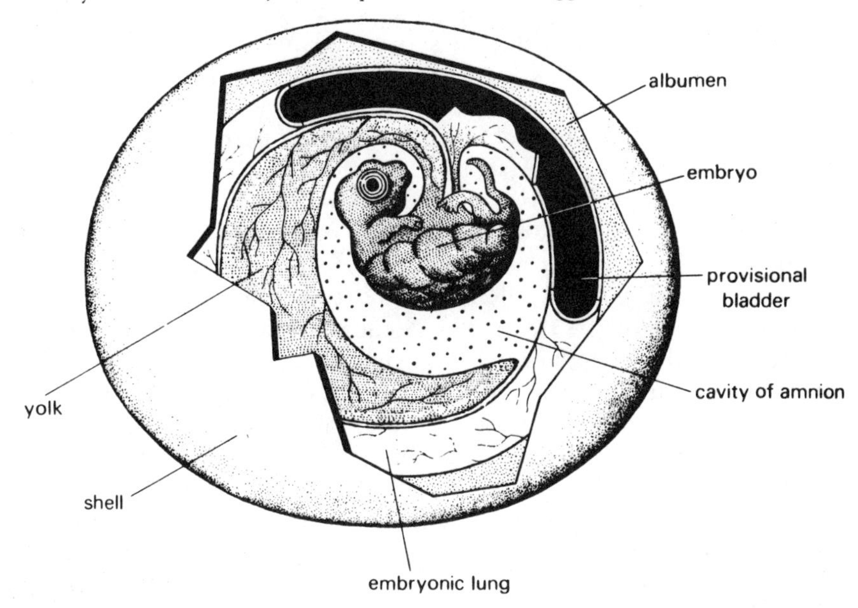

different from the egg of fishes. From the shell, constructed of crystals of hydroxyapatite and waxed over, to the altered chemistry, based on fat rather than protein, the amniote egg was in a different class altogether, a stunning advance on the simple blob of jelly that constituted the egg of frogs and fishes – a saltation if ever there was one.

As Professor Edwin Colbert comments; 'This was a major innovation in vertebrate history, to be compared with the appearance of the lower jaw, or the migration of the backboned animals from the water on to the land, and, like these preceding evolutionary events of great moment, the perfection of the amniote egg opened new areas for the development of animals with backbones.' It liberated the land-living vertebrates from dependence on water. No longer did they have to return to sea or river to lay their eggs, for their eggs had their own private pools of liquid to float in. So they could wander freely over the earth.

How and when did the change occur? We do not know. As usual, the fossil record is blank just when we most need it. The first known fossilised amniote egg comes from the Lower Permian sediments in North America, long after the reptiles had become established. It was about the size of a hen's egg.

North of the little town of Seymour, Texas, the Permian beds are well exposed and it is here that paleontologists found the skeletons of a creature almost exactly intermediate in structure between amphibians and reptiles. They called it *Seymouria*. Its skull is amphibian, its skeleton reptilian; in particular it has the typically reptilian number of bones in its feet; that is, two phalanges in the thumb, then 3,4,5,3 in the fingers as you pass from index to little finger. The strata are much too late for it really to be an ancestor of the reptiles – it must be a descendant of such ancestors. The thing we want to know is: did it lay eggs? Alas, there is no reply.

All we have to go on is the fact that very early in the story the reptiles diversified into an extraordinary number of groups, allegedly because the ability to lay eggs and move from the sea opened new niches to them. Some would go on to become the dinosaurs, others the lizards, yet others the turtles. Some would take to the air, some would lose their legs and become snakes. Yet others would move towards becoming mammals. But, what is rather astonishing, some went back into the water, like the crocodiles and, even more puzzling, into the sea to become very like fishes again. For whereas the plesiosaur was quite reptilian in appearance, with its long neck and paddles, rather than fins, the ichthyosaur was a pretty good imitation of a fish. Thus the argument that the reptiles succeeded so well because the amniote egg opened new possibilities of finding a living

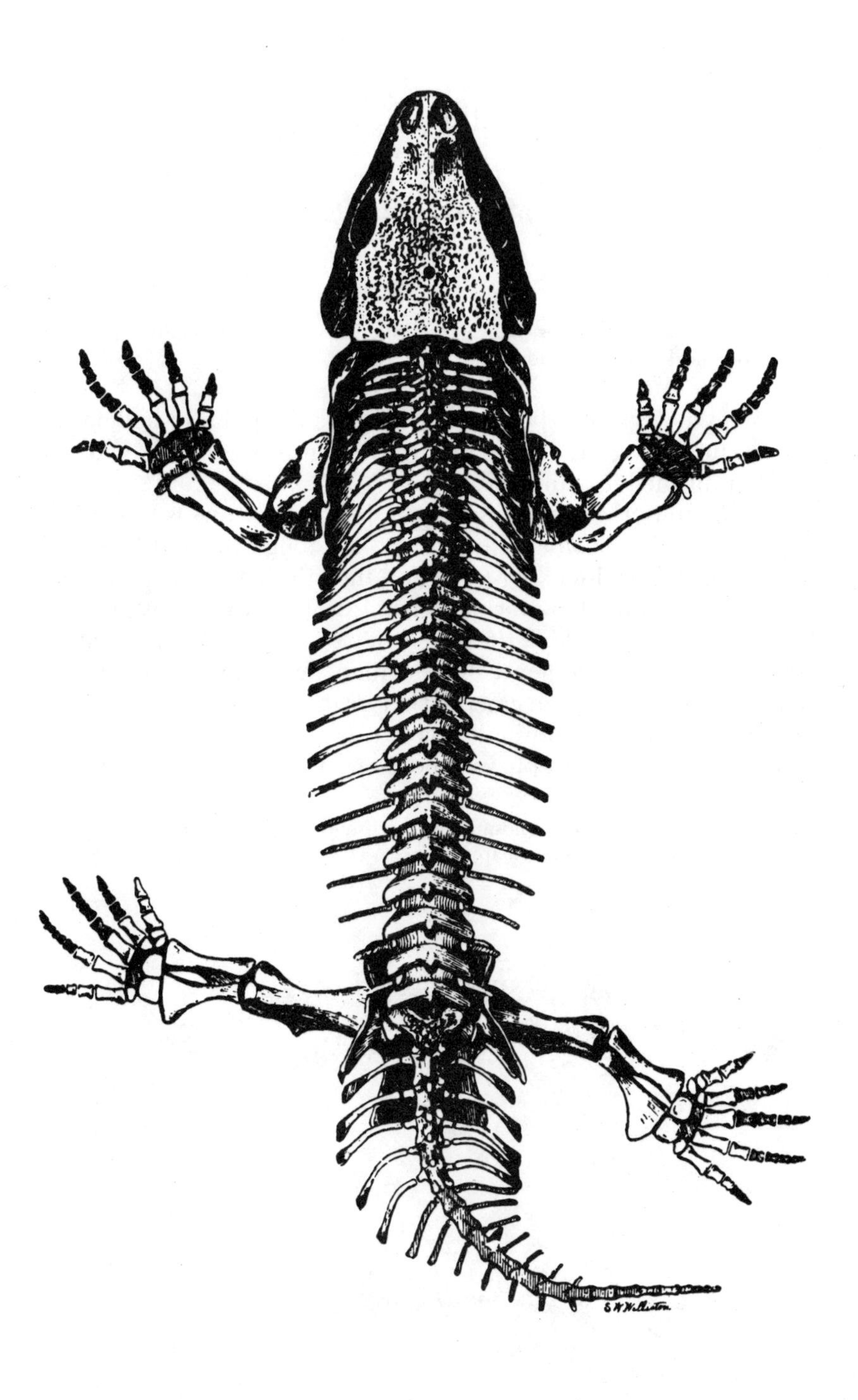

Skeleton of *Seymouria*, an intermediate creature between amphibians and reptiles

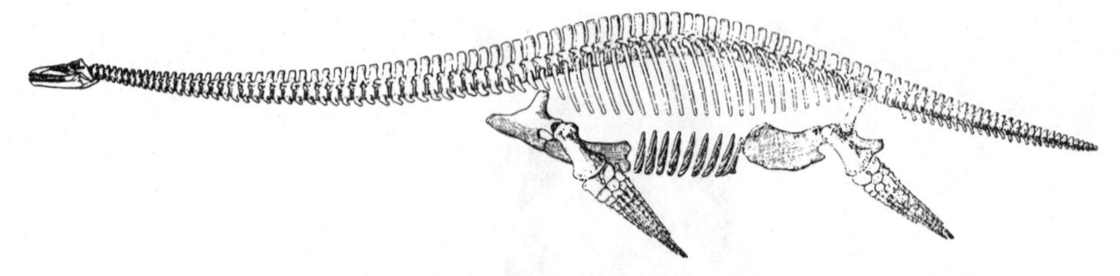

Muraenosaurus, a Jurassic plesiosaur. Its reptilian structure is especially conspicuous in the long neck and large paddles

begins to look extremely shaky. What was the point of having this wonderful new gimmick, if you were going to compete with the well-established fish anyway?

As a matter of fact, the ancestry of the ichthyosaurs is 'something of a mystery,' as Colbert says. They had long tails and long jaws equipped with evil-looking teeth and though they were quite small the world is probably well rid of them. Another mystery is provided by the mesosaurs, which do not seem to have left the water at all, showing that the emergence of the amniote egg was not an 'adaptation' to living on land but emerged before there was any need for it – a pre-adaptation, as evolutionists say.

There is no need to pursue the many ramifications of reptilian

By comparison with the plesiosaurs, the ichthyosaur's skeleton (top), like the dolphin's, is far more fish-like in appearance

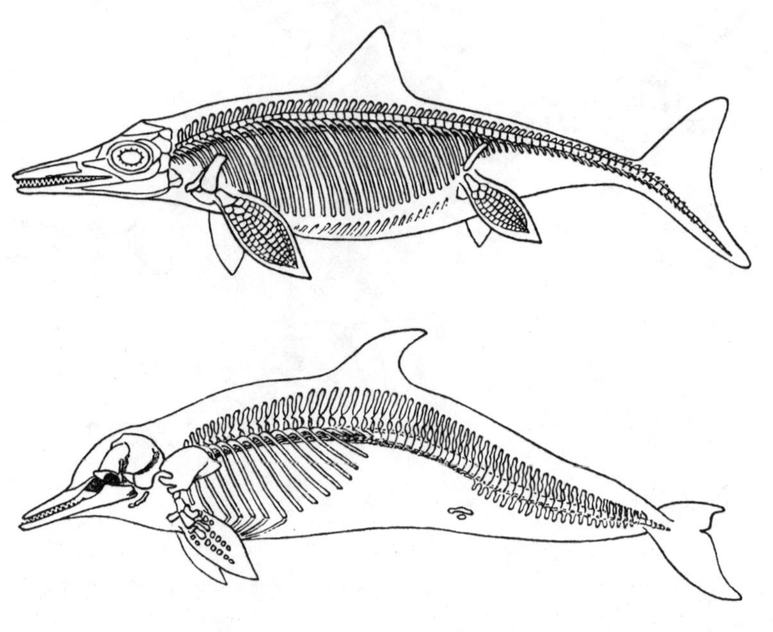

lifestyle, but I must enlarge on one which is important to my theme since it exemplifies another of these unexplained saltations in a particularly striking way. I am thinking of the jump from land to air and the origin of birds. The birds were not the first to take to the air, to be sure. The flying lizards had managed a clumsy sort of flight by spreading a sail of skin from an immensely elongated forefinger to their hind legs. But the birds achieved a degree of modification so amazing that it is hard to understand how they evolved at all.

3 *Learning To Fly*

What an extraordinary thing is a feather! So light and yet so strong. Under the microscope it is seen to be even more remarkable. The vane is divided into innumerable hollow 'barbs' each fringed with 'barbules'. Under higher magnification, each barbule is found to be equipped with hooks. These catch in the barbules forward of them, so that the whole structure is linked into a single vane, resistant to the air. Actually, feathers come in two models. The downy feathers which are designed to conserve heat (Designed? Perish the thought:

Stages in the development of the feather: tail of archaeopteryx (left) compared to the tail of a modern bird

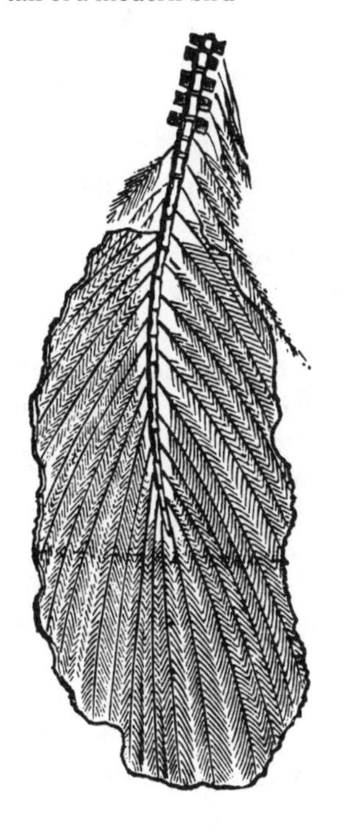

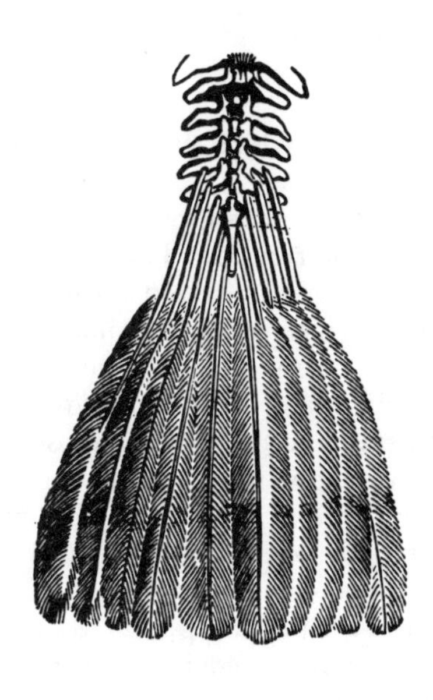

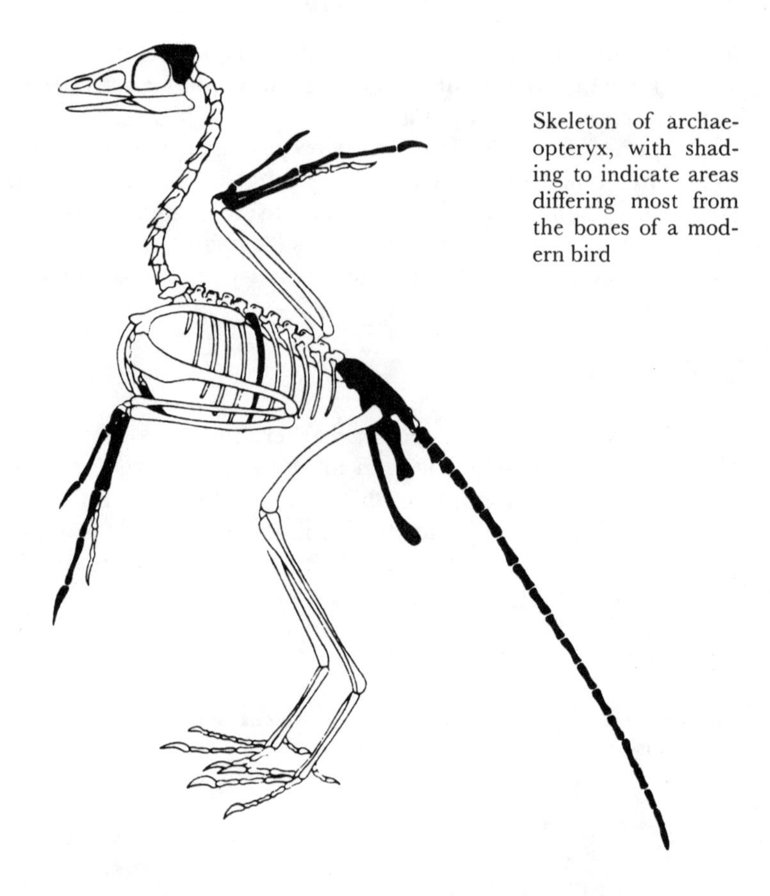

Skeleton of archaeopteryx, with shading to indicate areas differing most from the bones of a modern bird

let us say 'which by pure chance serve to conserve heat') lack the hooks.

Feathers are supposed by paleontologists to have evolved from the scales of reptiles. This, as evolutionist Barbara Stahl remarks, 'would have required an immense period of time and involved a series of intermediate structures. So far the fossil record does not bear out that supposition.' The paleontologists base their conclusion on the slight facts that both feathers and scales develop from papillae on the skin and both are composed largely of keratin.

We know little of the evolution of the first birds because fossils of them are extremely rare. The bird which died was, one may suppose, eaten by other animals or by ants, crushed by the feet of larger animals or the mere weight of vegetation. Such fossils as have been found are mostly splintered into fragments.

But paleontologists had one colossal stroke of luck. In 1861, men working in a quarry near Pappenheim, in Bavaria, came upon the skeleton of a creature which had been buried, with wings outspread,

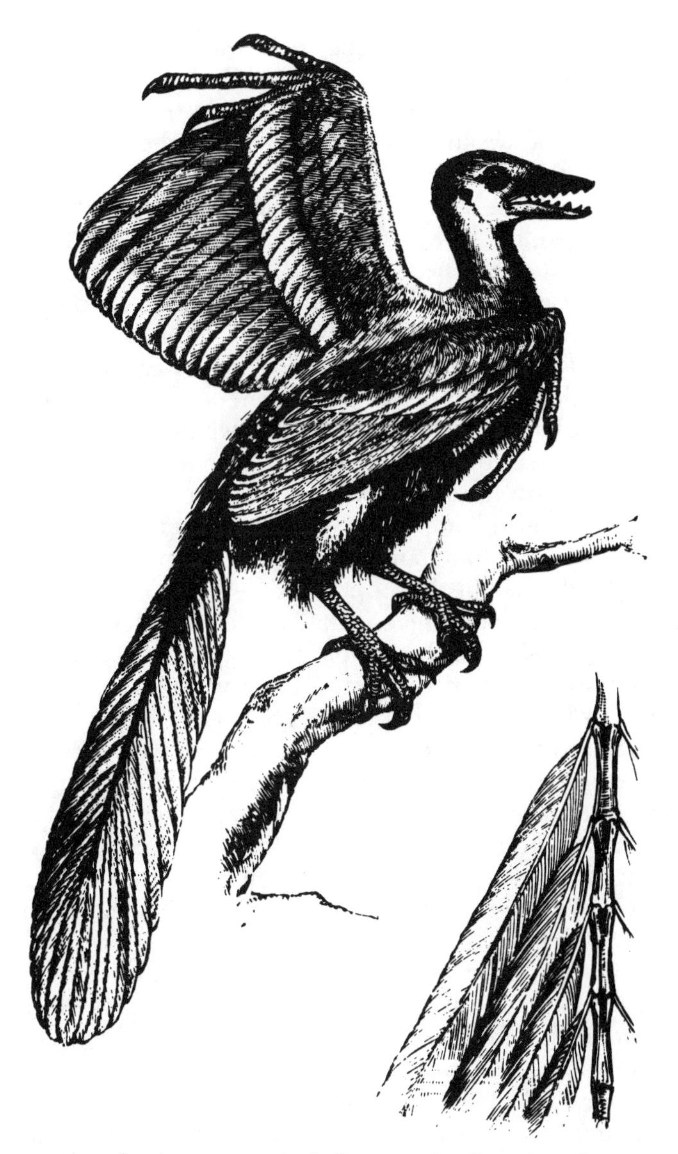

A reconstruction of archaeopteryx, including a section from the tail

in the sediments of the shallow seas which once covered the region. 'Recognising the fossil for the prize it was, they brought it to the medical officer of the district, Dr Friedrich Karl Haeberlein, who had an extensive collection of plant and animal remains, discovered in local quarries. He sold it to the British Museum where it was examined by Richard Owen and found to be missing only a few cervical vertebrae, the right foot and the lower jaw,' says Barbara

Stahl. Subsequently a second specimen was found about ten miles from Pappenheim and now rests in Berlin's Natural History Museum. This was archaeopteryx, a dyed-in-the-wool precursor of the birds. All that we know of bird evolution in the Jurassic has been gleaned from these two finds.

No one has seriously challenged the conclusion that the archaeopteryx was a bird, because its feathers left their imprint clearly in the rock and can be seen to be exactly like modern feathers, even though the creature itself has the long tooth-equipped jaws and trailing tail of a reptile. Closer examination also reveals reptilian features in the skull, pelvis and elsewhere. In particular archaeopteryx lacks the strongly developed sternum or breastbone serving as the anchorage of the powerful muscles which operate the wings in birds: hence some investigators have concluded that it could not fly – at best it could flap its way up to a branch to avoid a predator.

The creature also had other handicaps: it must have been heavy for its size, while its hand and wrist bones had not yet fused into an inflexible blade as in birds.

The astonishing point here is that the feathers seem thus to have evolved *before* flight and thus before they offered an advantage upon which natural selection could act. However, this is not a unique instance of such anticipation, as we shall see. Some commentators have suggested that feathers were first evolved to conserve heat; but the feathers of archaeopteryx are quite visibly the stiff flight feathers and not the downy heat-conserving feathers which lie beneath.

In point of fact, the number of modifications in reptilian structure which the birds have managed to effect in order to adapt themselves for flight is so large as to constitute a real problem and deserves our further attention.

To begin with, many modifications serve to reduce its weight. The bones are hollow, the skull very thin. It has abandoned the heavy tooth-studded jaw for the light but rigid beak. The body is condensed into a compact shape, the reptilian tail being abandoned, as also the reptilian snout. The centre of gravity has been lowered by placing the chief muscles beneath the main structure. Where organs are paired, like the kidney, and the ovary, one has been sacrificed. The pelvis has been strengthened to absorb (allow me the teleology) the shock of landing. The legs and feet have been reduced to minimum, the muscles operating them have vanished to be replaced by muscles within the body. The brain has been modified: a larger cerebellum to handle problems of balance and co-ordination, a larger visual cortex now that vision has become more important than smell.

Less obvious but even more remarkable is the change in bodily metabolism. To produce the energy for flight the bird must consume a lot of fuel and maintain a high temperature. Not only do birds eat a lot, as anyone who grows fruit or has seen the bullfinches systematically remove every bud from a treasured shrub knows, but they have a crop in which they can store reserve fuel. So that it can handle more blood, the partitions in the heart have been completed. The lungs too have not only been enlarged but are supplemented by air-spaces within the body. In land creatures like ourselves, much of the air in the lungs remains static; we exchange only a very small proportion of it in a normal breath. The bird, by passing the inspired air right through the lung into the air-sacs, contrives to exchange the lot with each breath. This system also serves to dissipate the heat generated by the muscles during flight.

It strains the imagination to visualise so many beautifully apt changes occurring by chance, even when one considers that 150 million years elapsed between the emergence of life from the sea and the appearance of the first birds. For my part I can imagine that each change might have occurred by chance during that time; what I find hard to swallow is the accumulation of different changes integrated into a single functional pattern.

Of these changes the most radical and the most baffling is probably thermo-regulation – the ability to maintain a high temperature. This was no doubt especially valuable to birds which soar into cooler air as well as having to remain mobile in winter conditions. We can see that feathers of the downy kind by insulating the body played their part in conserving the heat generated, but how temperature control gradually manifested itself remains a complete mystery. At one time biologists hoped that they might find different mechanisms or degrees of efficiency in temperature regulation in modern birds which would provide a clue as to its evolution, but no. Modern birds are all equally efficient in this respect. Thermo-regulation appears, full fledged, on the evolutionary scene with as little warning as the devil in pantomime.

If this were a biological treatise, I should now tell you something about the sea-birds and divers, mainly because a good deal is known. Such birds, dying and sinking into mud or sand, have been preserved more numerously than land birds. On the other hand I should pass discreetly over the forest birds and song-birds, of which nothing is known, because they have not. I cannot, however, altogether resist the temptation to mention the huge flightless birds, of which the king was the *Aepyornis* of Madagascar, the eggs of which are probably the largest ever seen, and which only became extinct in the last century. We still have, of course, the emu, the cassowary, the rhea, the kiwi, the ostrich and other flightless birds. In the Eocene, much larger and probably fiercer flightless birds existed, such as *Diatryma*, which stood about seven foot high and sported a tremendous parrot-like, scissor-sharp beak. Remains of *Diatryma* have been found in North America and Europe, while Europe also harboured a huge flightless goose-like bird, known as *Gastornis*.

The problem which these weird creatures, so much larger than any living bird, pose is: were they descended from flying birds which gradually lost the capacity for flight or are they birds which never

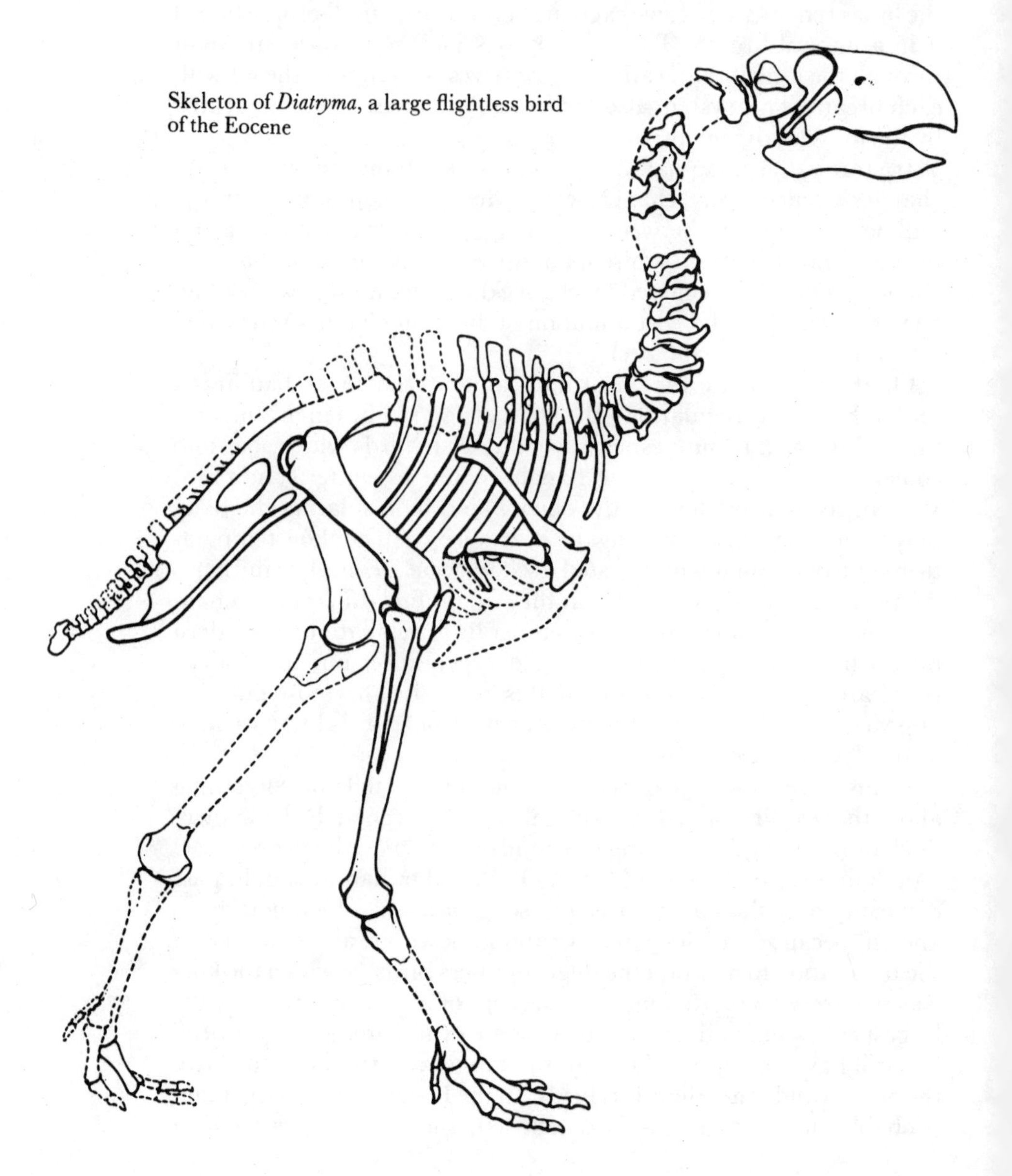

Skeleton of *Diatryma*, a large flightless bird of the Eocene

learned to fly? Whichever alternative you prefer, the implications for natural selection are uncomfortable. On the one hand, if the ability to fly bestowed an advantage on which selection could act, loss of that advantage must surely be a handicap which would soon eliminate the species. On the other hand, if flight bestows an advantage, why did birds which had so nearly made the necessary adaptation not finish the job? Had they perhaps got too large to fly? If so, why were not the genes which promote size winnowed out by selection? However you look at it, the flightless birds, the moa, the *Aepyornis*, *Diatryma* and the comic *Phororhacos* of South America, stand out like pustules on the fair face of the theory of natural selection.

Reconstruction of *Phororhacos*, a Miocene bird standing about six feet tall

Controversy surrounds the evolutionary origin of the penguins, but the general view is that they once could fly and swim, but lost the power of flight. No Antarctic fossil penguins have been found, but there are plenty in Australia, New Zealand and Patagonia, going back to the Eocene; unfortunately they are very like modern penguins and tell us nothing about their origins. (Penguins which have lost their flight feathers can still swim well: they fly under water really.)

4 *Reptile to Mammal*

The beginning of the mammalian story is rather different and even more obscure than those just described. More than a dozen different groups appeared simultaneously in the fossil record at the start of the Eocene, some 50 million years ago. Moreover they appeared in many different environments – in Asia, in South Africa and in South America. From that time, almost all the familiar modern groups – or, rather, their ancestors – from rabbits to tigers, from anteaters to bats, are rather common. There are but few fossils in the strata preceding the Eocene, but odd finds, which can be assigned to various dates spread over the preceding 150 million years, indicate that various preliminary attempts at mammal-making must have been going on. Four orders of mammals are inferred to have existed in the Jurassic, a conclusion based on a few teeth and fragments of fractured skulls. One of these orders, the Pantotheria, is claimed by some paleontologists to be ancestral to many modern mammals, a claim based largely on certain similarities in the teeth. Possibly, however, more than one group of mammal-like reptiles contributed to the mammalian stock, which would explain the sudden appearance of so many new lines. It is a puzzle unlikely ever to be solved.

What is odder, however, is that these several lines, although occupying very different niches, modified themselves in similar directions at much the same times. E. C. Olson, who investigated this thirty years ago, concluded that there must be some general kind of improvement unrelated to adaptation, a conclusion hard to reconcile with Darwinism. This is a point to which I shall recur.

There were, it is perfectly true, several mammal-like reptiles – so adjudged mainly because their teeth showed mammalian characteristics – and one group of these, the ictidosaurs, are said 'to have closed the gap between mammals and reptiles'. Their legs were pulled in under the body, unlike the sprawling reptile leg, and the skeleton was getting like that of mammals. Some of the theriodonts, for instance, had two phalanges in the thumb and big toe, as mammals do, rather than three or four as in reptiles. But were they similar in other respects?

Mammals have many characteristic features lacking in reptiles: a four-chambered heart and more or less constant body temperature; an insulating covering of hair, a single nasal orifice. The young are born alive (except in the curious Australian monotremes, like the duck-billed platypus). The brain case is much larger, the pelvic bones are fused into a single structure. And perhaps the most remarkable transformation of all: the bones which link the skull to the jaw in reptiles have retreated into the middle ear to become the ear ossicles – a switchover I'll come back to shortly.

None of these were present in the mammal-like reptiles. So something fairly drastic went on in the Mesozoic period for all these (and a few other) changes to have occurred within a geologically short time. It can reasonably be regarded as a saltation.

And it was at this time that the placentals and marsupials appeared. The marsupials are believed to have spread all over the world (though fossils are strangely few), but the great evolutionary burst of placentals at the start of the Cenozoic extinguished the marsupials in many areas. In Australia, which became separated from the other continents about this time, they lived on undisturbed, and in South America, which became separated from North America by the breaking of the isthmus in the Tertiary, much the same thing happened.

Marsupials, of course, are nurtured by the mother in a special pouch after birth and have various skeletal changes, notably a small brain case, which perhaps accounts for their defeat by the placentals. The American opossum is a 'living fossil' which tells us much about early marsupials. But in Australia there are no fossils of marsupials to go on before more recent times.

Meanwhile no fewer than twenty-eight orders of placental mammals evolved, sixteen of them persisting to the present time. Incidentally, the earliest mammals were insect-eaters and from them all the others, including ourselves, probably evolved.

Looking back, we see that in all four of the transformations I have described, massive changes occurred which the available fossils fail to record in detail. Thus the case for saltations, and against Darwin's slow accumulation of imperceptible changes, is strong.

5 The Flowering Plants

The flowering plants of our gardens, the grasses and cereals, all the well-known trees, from beech to palm, even the humble potato and

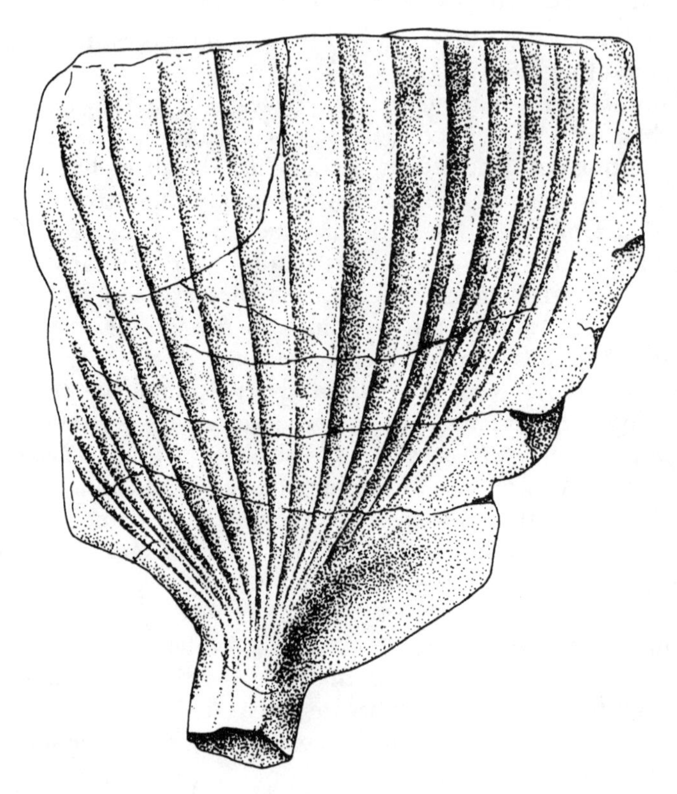

Fossil of the earliest known angiosperm leaf

cabbage, belong to a great super-order known as the angiosperms, distinguished primarily by the seemingly trivial fact that the ovules (from which the seeds develop) are enclosed in a neat case known as the carpel, as most gardeners will have noticed. This distinguishes them from the gymnosperms (the conifers and their allies), in which the seeds are unprotected.

But when did the angiosperms appear on the scene, and where? And from what group did they evolve?

Though there are plenty of theories, no one really knows. As botanist Tom Harris told the British Association in 1960 'I ask you to look back on an unbroken record of failure.' The University of Michigan's Charles B. Beck, while claiming that 'substantial and exciting progress' has been made since then, concedes that the mystery is still 'as fascinating today as when Darwin emphasised the situation in 1879'.

In the 'thirties most people assigned the origin of the angiosperms to the late Paleozoic, some 250 million years ago. Now most botanists favour the idea that they arose in the Cretaceous, a mere 120 million years ago, while others plump for earlier in the Mesozoic, say 170 million years ago. Arguments for an early origin were based on the presumption that a long time was needed for them to evolve, and the first undisputed angiosperm fossils are fully developed. But today it is beginning to be accepted that evolution can be quite rapid. The reason no one has found any fossil flowering plants in the Cretaceous is simply that they haven't looked, says N. F. Hughes of Cambridge's Sedgwick Museum. However, just to confuse the issue, the biochemists have been analysing the amino-acid content and declare that they must have originated 400 to 500 million years ago. Yer pays yer money and yer takes yer choice.

And where did they come from? The popular notion is that they came from the tropics, for relatively early angiosperm fossils are found all over the south-west Pacific and south-west Asia. But in the past decade or two geologists have come to accept the idea that the earth's surface consists of giant plates floating on a molten sea of rock. And that the land mass was once all one, and broke up, the Americas detaching from Africa on one side, the Australian and Indian continents on the other. So everything depends on *when* they originated if you want to calculate *where* they were, in global terms, at the time.

There is also a sharp split between those who (like Professor G. Ledyard Stebbins) think they originated as small bushes in the uplands, and those (like England's N. F. Hughes) who declare that upland plants never colonise lowland areas, but rather the other way about.

And what was their ancestry? The obvious place to look is among the gymnosperms or conifers, but no satisfactory intermediate forms between this type of seed mechanism and the flowering-plant kind have been found. The next place to look is among the ferns or *Pteridophytes*. But the evidence is just as mute or moot here. Beck sums it up bluntly. 'We do not yet know when the angiosperms originated, although we may be getting close to an answer. We do not know exactly where they originated. We are even less certain of the ancestral group from which they originated.' But then he adds doggedly: 'Nevertheless, I do not entirely agree with Tom Harris's pessimistic statement of 1960. Certainly there have been many failures . . . but I am optimistic about future successes . . .'

6 Summing Up

Looking back over the evidence, we see that in each of the major steps there is an almost complete absence of fossils capable of supporting the claim that the new forms arose by the gradual accumulation of minute changes. Eggs are first found fully developed; so are feathers. Fossils of early birds are 'extremely rare'. Placental mammals appear simultaneously in twelve groups. Some authorities would put the figure higher; actually, there are some twenty-six groups of mammals the origins of which are completely obscure. As for the fishes, 'we do not know the direct ancestors of the fish,' says Professor Lehman. The origin of the insects is a complete mystery. The position as regards the flowering plants is no better.

Professor G. G. Simpson is an ardent Darwinist, but he goes so far as to say: 'the absence of transitional forms is an almost universal phenomenon'. This is true of invertebrates as well as vertebrates and also of plants. He adds: 'The line making connection with common ancestry is not known even in one instance.' The rodents, he notes, appear suddenly, already equipped with their specialised gnawing teeth. As to the mammals, 'In all 32 orders of mammals, the break is so sharp and the gap so large that the origin of the order is speculative and much disputed.'

Even the more recent evolutionary changes are baffling. For instance, the whales and dolphins appear in the fossil record fully formed and diversified. We do not even know what mammalian order they sprang from. Again, in South America, there are puzzles such as those strange and rare creatures the Groeberidae, with their unique teeth. 'These highly specialised animals appear in the fossil record without any antecedents,' complains Simpson, 'and then immediately – geologically speaking – disappear again for ever.'

The problem reaches its acutest form when he considers not classes or orders but whole phyla. Thus an enigmatic group of creatures emerged in the Cretaceous, known, from the unusual nature of their joints, as Condylarths. They are believed to have given rise in due course to horses, camels, elephants and other creatures known today, though the links, as usual, are missing. But where did the Condylarths come from?

Paleontology throws no light on the origin of the phyla, is Professor P. P. Grassé's conclusion. As editor of the twenty-five volumes of the *Traité de Zoologie* he ought to know.

So it is rather remarkable that Darwinians can dogmatically deny the occurrence of saltations and continue to assert that these major changes arose by the accumulation of minute differences. (Simpson tries to dismiss the sudden appearance of new taxa as 'artifacts of sampling and the taxonomic system', an argument so feeble as not to be worth examining.)

Recently one or two biologists have attempted a compromise. On the whole evolution took place gradually, they say, but this trend was punctuated every now and then by a sudden jump. This is known, accordingly, as 'punctuation theory'. The most prominent of these is S. M. Stanley who, writing in 1979, points out that paleontologists tend to look at large changes in large populations and biologists at small changes in small populations.

In this development there is a certain irony, for as long ago as 1900 there were hot debates between the Darwinian gradualists and the sudden mutationists. In the 'forties, Richard Goldschmidt again challenged the orthodox, but Professor Sewall Wright, the eminent American geneticist, said that evolution was rapid in small, isolated groups but not elsewhere. Mayr opposed this and in 1954 put forward a punctuationist view. Stephen Jay Gould of Harvard and Miles Eldredge of the American Museum of Natural History, New York, restated it in 1972, only to be challenged by C. W. Harper in 1976.

The official position remains that these huge changes occurred by rapid evolution in small groups – so rapid and so small that no fossils were left. While this is just possible, it is no more than an educated guess, and more than one biologist has found it slightly improbable. The trouble with small groups is that they die out more easily than large groups, so that any major alteration which unfits them for the niche they are in is liable to eliminate them before they can find a new one.

In the seventeenth century the British naturalist John Ray said that no species ever gave birth to another species, meaning that rabbits do not give birth to hares, nor owls to pigeons. That observation remains true. The changes that occurred certainly occurred over many generations. A million years is a brief time geologically but it could encompass anything from 200,000 up to a million successive generations for the kind of creatures we have been considering. The question at issue is: not whether rapid changes occurred but did those changes take place as a result of natural selection or something else?

So far I have discussed, almost exclusively, the main changes in ground plan during evolution. But within these large changes occurred some very significant innovations, notably the creation of the eye and the ear. There are other puzzling features too, when we survey the whole sequence spread out over time: speedings up and slowings down, the vanishing of novelty as well as its appearance. These topics form the substance of the next chapter.

CHAPTER FOUR

By Fits and Starts

1 Has Evolution Stopped?

School biology leaves many of us with the impression that evolution marched steadily and implacably onward and upward. Such is not the case. Many strange alterations of rate occurred and are essentially unexplained, though many guesses have been made.

Back in the 'thirties, the eminent physicist Professor J. D. Bernal advanced the idea that evolution was steadily accelerating. It took about a billion years for the first life forms, such as bacteria, to appear, about half a billion more to produce the filamentous algae, and as much again to produce the eukaryotes or nucleated cells. The first fish appeared 500 million years ago, the first reptiles, 300 million years ago, the first mammals, 200 million years ago. Thus the intervals shortened.

But to say evolution is accelerating means, presumably, not just that the really major changes occurred at shorter intervals but also that novel forms within large groups were appearing increasingly frequently.

However, exactly the contrary may be true.

In contrast with Bernal, Professor C. D. Darlington, Sheridan Professor of Botany at Oxford and one time director of the John Innes Horticultural Institute, considers that evolution is slowing down. He points out that the primitive organisms known as prokaryotes evolved early and then settled down: they do not seem to have changed appreciably in the last billion years. Professor Grassé is another who thinks evolution is slowing down or even coming to a stop. So is Professor James Brough, Professor of Zoology at Cardiff University. He points out that no new phyla have emerged since the Cambrian age, 500 million years ago. Since then evolution has been restricted to working within about a dozen different patterns. Moreover, the emergence of new classes within phyla had ceased by the Lower Paleozoic, around 400 million years ago. When we descend to the next taxonomic category, the orders, we find that of forty-seven known fossil orders forty had evolved by that time; in the next fifty million years (the Devonian) only three more appeared; and in the next 170 million (the whole Mesozoic) only four, since when none have. There has also been a marked slowing down, Brough holds, in the production of new families. 'As to the future,' he concludes ominously, 'evolution may go on working in smaller and smaller fields until it ceases altogether.'

These are fighting words of course, for the orthodox geneticist assumes that the rate of mutation stays constant unless it can be shown that some unusual environmental effect has increased it. On a local scale, to be sure, an increase can easily happen, for example if a rise occurs in the level of ionising radiation. But for mutation rates all over the planet to be affected simultaneously, we should have to assume that the solar system, in its travels round the galactic centre, had entered areas of intensified cosmic radiation, or some titanic event of that character.

Brough does not stick his neck out by speculating on these lines. He simply says that evolution in the more distant past was different in some way from evolution in the present. Was the evolution rate fantastically high, he asks. If so, this could explain the failure of paleontologists to find any missing links. This is, of course, anathema to the old guard.

These are very puzzling facts and pose great difficulties for the theory of natural selection. If Darwin was right in thinking that the small variations led in time to the larger ones, we might expect new orders, new classes and new phyla to form with increasing frequency. Divergence would become steadily greater. Instead, we get a branching structure rather like that of a beech tree. A few main branches are formed. These in turn divide into smaller branches but the number of main branches remains constant. Similarly the smaller branches subdivide but do not increase in number. To complete the analogy, we could add that quite a few main branches have dropped off, and so have many lesser ones. A tree grows in this way because, in the older wood, certain processes of subdivision cease to function. The inference for evolution is that there were diversifying forces at work in early times which today are only active at lower levels. At the genetic level, this suggests that parts of the genome become insulated from or unresponsive to the forces which cause variation.

Reflecting on this, the thought occurs that a tree finally ceases to make significant growth, its general form remains stable. Does evolution also tend to a final state, marked only by trifling losses and gains at the species level?

But there is a further anomaly to account for: the existence of many genera which contain only one species, and of families which contain only one genus. For if species are constantly diverging, we should expect each taxon to have many members, just as a tree bears many branches. If we saw a tree, many of the branches of which bore only one branchlet, and many of the branchlets only one twig, and this twig only one leaf, we should be amazed and suspect the presence of some strange disease.

Dr J. C. Willis, the botanist who worked in Ceylon, whom I have already mentioned, investigated this question; what he found was as puzzling as it was surprising. Plotting the number of genera against the number of species, he got a family of curves. Thus the plant family known as Siparune ranged from one genus with thirty species down to twelve genera with one species each. In other words, the power of diversifying seems to be distributed unequally between genera in a statistically regular way. Willis pointed out that this kind of distribution is found in quite other fields. For instance, a town will usually contain one de luxe hotel, several not quite so fancy and so on down to a large number of boarding houses and even more bed-and-breakfast places. The same sort of thing is true of surnames. Some names are very common, others less so, and so on down to names which occur only once. As Willis sarcastically points out, this kind of distribution is not produced because some hotels, or some names, have more 'survival value' than others.

Willis also noticed another curious fact, which could be connected with all this. Just to take an example, the cabbage-like plant *Coleus barbatus* is widely distributed in Asia and Africa, including those parts of Ceylon where he was working. But its near cousin *Coleus elongatus* is known only in Ceylon. Willis found many other instances of this puzzling distribution and found it impossible to believe that natural selection could so have favoured one species but not another barely distinguishable from it. No doubt between these extremes were other members of the family with varying distributions, giving the same kind of statistical pattern as before, but this he did not have the means to investigate. The explanation he advanced was that the successful species were simply those that got there first, just as the first firm to enter a new market tends to mop up most of the available business, leaving only some specialised variants to be supplied by latecomers.

Darwin noted that the most widely distributed families are the least complex ones. Willis's explanation accounts for this rather neatly. What it fails to explain is the occurrence of so many extinctions. Could it be that species have a natural 'life-span'? It has been calculated that the average 'life' of a genus is twenty-four million years and for species one million. But whether this figure is meaningful or merely the average of a number of chance events it is hard to say. There seem to be signs of senescence and disorganisation in the shells of ammonites towards the end of their reign. Originally these were straight structures which grew by adding new sections. Then by growing faster on one side than the other they became curved and finally wound themselves into tight whorls. After millions of years in this form they began to uncoil and show other defects of development such as irregular coiling, and soon vanished. However, the question of extinctions is a complex one which we are going to look at more closely before the end of this chapter.

Perhaps the question which people like Willis have raised, 'Does evolution proceed from the top down or the bottom up?' is wrongly posed. Darwin generalised his guess into too sweeping a form. It looks as if evolution developed in two phases, an opening one in which certain major patterns were established and a second phase in which speciation worked within narrower and narrower set limits. We shall come back to this earlier phase in another chapter.

2 *Explosive Radiations*

One of the most astonishing features in the fossil record is the way in which new phyla have quietly appeared and carried on without making much impact for a while, and then have suddenly diversified into numerous different forms. This is called by paleontologists 'explosive radiation'. (No connection, of course, with the radiation from an atomic explosion, or even from a radiator. The word is used merely in its basic sense of lines radiating from a point.)

The mammals are as good an example as any of explosive radiation. Though they showed up some 200 million years ago – as we judge from scattered instances of forms long extinct, like *Eozostrodon* and the Symmetrodonta – they only began to ramify seriously about 75 million years ago, at the start of the Paleocene. (Even this assertion is only an inference as regards most lines, since the first fossil forms only appear at the *end* of the Paleocene.) Most mammals arose, it would seem, in the remarkably short space of 12 million years. Suddenly we find remains of carnivores, of cetaceans (whales, dolphins), of rodents, of marsupials, of toothless creatures like the anteaters, of horses, camels, elephants, rabbits, bats and many others. The great majority of these new mammalian forms are still extant. Only a few versions, such as the Creodonts and the Desmostyles, gave up the struggle to survive.

With reptiles the story is a little more complex. There was an initial radiation at the start of the Triassic which produced a lot of forms, all of which failed towards the end of the Cretaceous (except the crocodiles); plus three or four radiations in the geological period following, the Jurassic. The finding of various extinct reptilian forms also implies an even earlier radiation two geological periods before the Triassic, when the amphibians first gave rise to reptiles. Or take the fishes: they show two radiations. In contrast, the plants seem to have got going and never looked back.

What caused these mysterious explosions? Was it a change in the environment offering new possibilities of survival, or a change in the mutation rate creating new forms, or perhaps a combination of each of these? Or were rivals under pressure?

The first person to face the issue of tempo fairly and squarely was George Gaylord Simpson, whose distinguished career included being Curator of Vertebrate Paleontology at the American Museum of Natural History, and holding the Agassiz professorship at the Museum of Comparative Zoology at Harvard. His *Tempo and Mode in Evolution*, 1944, remains a classic.

When we analyse the situation more closely, he pointed out, we find some lines which evolve very slowly if at all, and others which evolve very rapidly, with a much larger group between the extremes. In this middle group there is a certain spread of rates, within each phylum, but the very slow and very fast are right outside this range.

Many bivalves, such as oysters and mussels, have evolved very slowly, changing little in 400 million years. Simpson says that, among them, 'I find no evidence that a bradytelic (slowly evolving) line has ever become extinct; they seem to be virtually immortal.' Other slow-evolving lines include the coelacanths, which have been around since the Devonian, the king-crabs unchanged since the Triassic, and the opossums. The list of these 'living fossils' could be made much longer. There are also lines which have remained unchanged in broad outline, or structural type, while diversifying in minor respects, such as the rabbits, bats and armadilloes.

For an example of fast evolution we need look no further than man himself.

Simpson thinks that the evidence suggests that the slow-moving lines were once fast-moving. It is as if some newly evolved form tried a number of variations, most of which failed, but one of which settled down comfortably in a niche. However we also find slow lines throwing off variants which in turn throw off still more specialised forms, and so on. After a few million years these begin to fail in reverse order, the last evolved vanishing first, until only the original line is left. But, as usual, there are no intermediate forms in the fossil record, so this is only an inference.

All this is, of course, very awkward for the theory of natural selection, which assumes that the fast-evolving types will survive because they can adapt themselves to changing circumstances, while the slow-evolvers will become outmoded and vanish. Just the reverse seems to be the case. But worse is to come. The accepted view is that rapid evolution is brought about by three things: a high mutation rate, a short time between generations, and a large population – since a large population is more likely to throw up favourable mutations and less likely to be extinguished by unfavourable ones. But man has or had none of these. The population of Europe in stone age times was probably only 100,000 and the age of sexual maturity much higher than today and certainly much greater than most other animals. The mutation rate was low. Conversely bacteria, which satisfy all three conditions, have evolved little. A theory which predicts results one hundred per cent contrary to the observed facts is clearly faulty.

When we look at particular genera, the question of evolutionary rate becomes, if possible, even more puzzling. For instance, our old friend the coelacanth started off with a period of rapid evolution and then settled down into complete stagnation. One could say much the same for the sharks and lampreys.

The long and short of it is, as Brough says, the origin of explosive radiation remains 'obscure'. Simpson attempts to explain the explosive radiations by supposing there was a period of environmental perturbation which forced numerous lines to adapt. This is a weak argument since it is more likely that many species, failing to adapt, would have perished. Moreover different species were affected at different times. And as we shall see in the next chapter, it is highly doubtful whether there was any link at all between *major* evolutionary changes and environment. Others, like Stanley, see in this proof of their claim that speciation occurs by a different process from phyletic (long-term) evolution. In their estimation, evolution consisted of slow development punctuated by sudden jumps (hence it has been called punctuation theory). It is a brave attempt to satisfy both parties.

Finally, while we are discussing the time factor, there is the question of whether there was enough time to allow for evolution. Or rather, since the evolution undoubtedly occurred, whether the theory holds. Darwin himself had doubts about this and so have modern authorities, such as Professor C. H. Waddington. One of the few facts we can turn to, in an attempt to answer such a question, is the opening up of South America to North American animals. About three million years ago land bridges formed between North and South America and a wide variety of animals streamed in to South America, where they displaced many of the indigenous species. In those three million years, quite a number of new species have arisen but no new genera. This rather suggests that some very much longer period is required to generate new orders and still longer for new phyla. Something exceptional is therefore required to explain the relatively rapid appearance of new orders and phyla which in fact occurred.

The causes of explosive radiation are, to put it bluntly, a complete mystery. Such explosions may be due to the opening of a new niche, to the elimination of rivals, or it may be a response to stress, a search for new modes of existence, when old ones become too strenuous. In short, it may be because things are too easy or because they are too tough. Who knows? Either way, it suggests that mutation rates can be stepped up or damped down according to need. This implies the existence of mechanisms of which Darwin did not dream.

One must not think of phyla as merely existing for millions of years and then expiring. Usually it is the case that they start from small beginnings, become more and more numerous, then fade away. Sometimes they start to fade away, then recover from a low point, and then maintain themselves modestly. Hence the problem of failure is coupled with the problem of success, which I discuss elsewhere.

3 *As Dead as the Dodo*

Natural selection not only brings new species into existence – if it does – but also eliminates species, and on a colossal scale. It is calculated that 99 per cent of all the species which have ever existed are now extinct. So perhaps it may be more instructive to discover why species vanish than why they appear.

The stock view is that organisms fail to adapt sufficiently fast to changing circumstances. Simpson says, for instance, that the lag between 'environmental demand and evolutionary response' is 'the usual or universal cause of extinction'. 'Usual' perhaps, but 'universal'? Let us see.

What are the changing circumstances which might have this drastic effect? First, there is a change of climate. This might not affect organisms directly but would probably affect their way of life and in particular their diet. One can imagine that the gradual retreat of the forests and the consequent extension of grasslands must have been hard on forest-living creatures which had to learn to graze or starve. Similarly, the drying out of marshes or conversely their creation. Again, there have been periods when the sea was much more salt, and others at which its calcium content was so low that the shells of marine creatures would dissolve.

However, another possibility is that a group might be too successful and consume all the creatures on which it depends for food. Population crashes for this reason are not uncommon today, and though normally the group recovers, in some cases the process might go too far, and the few remaining survivors fall to predators or fail to hold their own against competitors for food. Again, the appearance of a new or more efficient predator could extinguish all victims who could not run faster or devise better strategies of escape.

In all such cases, the danger is over-specialisation. The creature which mops up all its food supply will survive if it can learn to depend on some other kind of food, which may mean adapting its structure, as happened when browsers became grazers, and needed a different set of teeth.

For a supreme example of the dangers of specialisation, we may look at the green turtles of Florida, which swim out every year to an island in the Atlantic to lay their eggs. They will lay nowhere else. If the erosive action of the sea ever destroys this small island, they will be in danger of extinction. They have literally put all their eggs in one basket.

To blame over-specialisation in this airy way is all very well, but the charge raises various questions. *Why* does one group specialise and another not do so? As depicted by Darwin, adaptation is precisely a mechanism for specialising. It was precisely the fact that the beaks of his finches were specialised that aroused his interest. To avoid specialisation is thus to avoid the process of natural selection. The evolutionist who condemns over-specialisation is thus in conflict with himself. Moreover, if non-specialisation is advantageous, there should have been selection for a mechanism which ensures it. In short, Simpson's throw-away line doubly invalidates the Darwinian thesis.

There is a mass of paleontological evidence to show that it is the most highly adapted creatures which vanish; life continues by the evolution of less specialised ancestors. So it would seem that a high rate of evolution is undesirable. It is the creatures which evolve least which last longest. The microbes which, though they varied about a mean, never advanced are still with us after 3½ billion years. There are ten or so orders known collectively as Pelecypods – the oyster is a familiar example – which have survived since the Carboniferous or earlier with only minor modifications, a period of 200 million years or more. The slow evolvers 'are almost immortal', as Simpson says.

The curious thing is that Simpson does not seem to see the complete contradictions in what he says. He starts by asserting that extinction comes from not evolving fast enough; then he shows that it is precisely fast evolvers which die out. Since the second proposition is known to be true, it must be the first which is faulty. And in that case the theory of natural selection stands in serious need of revision.

Professor G. C. Williams of Princeton takes a completely contrary and unorthodox view. He declares that evolution rates are always quite high enough to adapt to new circumstances, and points to Darwin's own work on animal and plant breeding. Man has produced dogs as diverse as the Pekinese and the Great Dane within a few generations. We have already noted how rapidly the maize plants learnt to grow short or tall, and how swiftly Kettlewell's moths adapted to industrial conditions. The problem, as Williams sees it, is precisely to *prevent* evolution proceeding too fast and producing an excessive specialisation. Mutations, which orthodox biologists regard as the raw material of evolution, he would describe as functional mistakes. And genetics becomes the machinery of stability, not the machinery of change. How to resist change he suggests, is the real problem.

This startling reversal of received opinion makes better sense of the facts and calls for a lot of reconsideration of work done and theories advanced. But it leaves Darwinian selection in a worse case than before.

There are two other possible causes of extinction which deserve mention. First, the tendency of evolutionary trends to run too far, which I have earlier called 'overshoot'. I remind you of the sad case of the Irish elk with its unwieldy antlers. (The stock explanation, you will recall, was that the size of his horns was controlled by the same genes as controlled body weight, and a large body was advantageous. But this did not seem to work for *Gryphaea*.) Simpson rules 'momentum' out completely, saying magisterially: 'There is no good evidence that a trend has ever continued by momentum beyond a point of advantageous or selectively neutral modification or has ever been the direct cause of extinction.'

Second, a more recondite possibility, is the unbalancing of the sexes. Since one male can fertilise many females, predominantly female populations are possible and exist. In theory, the factor for maleness could be bred right out. However, most if not all such populations can revert to asexual reproduction. It seems highly unlikely that this was the cause of the majority of extinctions, even if it could be shown to account for a few, here and there.

4 *Catastrophe Theories*

There remains one long-standing and highly puzzling problem concerning extinction: the periodical occurrence of catastrophic extinctions in which a wide range of orders and even whole phyla are wiped out simultaneously. These major catastrophes seem to have occurred at the end of geological periods and are in fact used to date the latter. The most drastic of all these disasters was the one which occurred at the end of the Permian epoch, when 75 per cent of the amphibia and 50 per cent of the reptiles suddenly vanished. In all, twenty-four orders vanished, including the fusilinids, shelled creatures which had populated the oceans for eighty million years; their shells, sinking to the bottom, created thick deposits of limestone. It took the world more than fifteen million years to recover from this hecatomb.

The catastrophe at the end of the Cambrian period was almost equally disastrous: two-thirds of the sixty families of trilobites vanished. So was the catastrophe at the end of the Triassic, when twenty-four of twenty-five families of ammonites gave up the struggle. At the end of the Cretaceous about a quarter of known animal families succumbed: sixteen orders or super-families perished.

The fallen included marine reptiles, flying reptiles, the dinosaurs and even some classes of the marine creatures known as foraminifera and phytoplankton. Several types of fish gave up the struggle, as also many belemnites and ammonites, although bottom-dwelling fish, some mammals and crocodiles survived.

Unrelated groups in different habitats seem to have vanished at the same time, including both land and marine creatures, though, in general, plants were not much affected, except to some degree at the end of the Permian.

Sceptics point out that the absence of fossils does not necessarily mean that a group was extinct: conditions may just not have been right for fossilisation, or the strata containing them may have subsequently been destroyed. The remains of some supposedly extinct groups do reappear, proving that they were never extinct. A case in point is the discovery, in 1978, by Anne Warren of Latrobe University of a temnospondyl (a class of reptile) supposedly extinct some thirty million years earlier.

But the mass extinctions I have described cannot be explained away so easily: they were on too large a scale. Some great natural disaster must have occurred. Speculation runs riot. Could solar flares have destroyed the ozone in the atmosphere to the point at which ultra-violet light was admitted in large amounts? Or could the earth have passed through the debris left by a supernova, which might have had the same effect? But ultra-violet light should not have affected marine creatures.

The fact that major extinctions coincided with the end of geological epochs rather suggests that the cause was terrestrial and geological. One ingenious theory postulates that the flow of nutrients from the land into the sea declined as the continents were reduced in level (as occurred from time to time) thus starving the plankton which generate much of the oxygen in the atmosphere. The fall in oxygen and consequent rise in carbon dioxide made it too tough for the dinosaurs. The turtles and crocodiles survived by going into torpor. However, since the mammals survived it isn't very convincing.

Professor Norman Newell thinks that rising sea levels flooded great land areas and destroyed the plants on which many animals depend for food, but doesn't explain why this decimated the ammonites and marine forms.

In 1979, the prominent physicist Luis Alvarez came up with new evidence. As he is the assistant director of the Lawrence Radiation Laboratory, the holder of (among others) the Einstein medal, and the discoverer of tritium and various nuclear phenomena, his suggestion must be taken rather seriously. It began from the detection of unusually high levels of iridium near Gubbio in Italy. Subsequently four more instances were reported from different parts of Europe. Then osmium and iridium enrichment was reported from Spain. These elements are rare on earth and to account for these facts, Alvarez suggests that a large meteorite hit the earth at the end of the Cretaceous. If it fell in the ocean this would account for no crater having been formed. In addition to the heating effect, dust might shut off the sun's heat causing severe cooling for a while. And in fact the earth did begin to cool from this time.

As an alternative, Kenneth Hsu, who is head of the Geological Institute in Zürich, has suggested that a comet hit the earth, heating the atmosphere enough to extinguish the cold-blooded reptiles and releasing cyanide which killed the sea creatures, as well as causing a rise in carbon dioxide, which would raise the point at which the shells of sea creatures dissolve. It is true that deep-sea drilling has revealed remarkable changes in ocean chemistry at the time of the Cretaceous disaster, but in such a case it is hard to see why anything survived. And as there were at least six big extinctions, the earth would need to have been hit by six comets.

An argument against all such theories is the relative immunity of plants. However, in 1978 a Russian scientist claimed to have found an abrupt change in the flora at just this epoch.

A good deal depends on the very delicate dating techniques used to establish the age of strata, from which the date of extinction is deduced. One study shows the dinosaurs as vanishing half a million years later than the foraminifera, but another makes these events coincident. Again, the iridium measurement techniques are laborious and it will take time to discover how general the enrichment was. So the problem is likely to continue to intrigue us for some time.

Catastrophes apart, it is apparent that natural selection does not account for the extinction of large groups in any convincing way. Are there unknown stabilising mechanisms in genetics which get out of control and prevent adaptation? Or is the problem not one of structural adaptation at all? Was it some rigidity of *behaviour* which – as in the imagined case of the green turtles – spelled destruction?

As the great Ernst Mayr has admitted: 'The frequency of extinction is a great puzzle to me. Far too little attention has been given to the factors responsible for [it].'

5 *Unnatural Selection*

The facts we have been reviewing certainly pose considerable difficulties for the theory of natural selection.

If the mammals acquired their new features because these had superior survival value to the reptilian features, why did the mammals keep a low profile for 150 million years before beginning to supplant the reptiles? It has been asserted in many textbooks that being warm blooded, the mammals could move faster and prey on the smaller saurians. But they do not seem to have done so to any great effect, while the large saurians were protected by their size – and perhaps, in some cases, by their ability to retreat into deep water. Contrariwise, if the saurians were advantaged by their size, what was the evolutionary advantage in being a hot-blooded mammal?

This argument is brought into still greater confusion if the theory advanced by Yale's Robert Bakker and others is true, that the dinosaurs were actually warm blooded.

Then there is the other half of the story. Why, in the Cretaceous period, did the mammals suddenly become dominant? Was it because the dinosaurs died out and left the field free? But what could have hit the dinosaurs so hard without affecting the mammals? If it had been the kind of cosmic disaster I have been describing, it probably would have hit the mammals as hard as the reptiles.

Similar questions can be asked about plant evolution. The angiosperms or flowering plants took over from the gymnosperms (conifers, etc.) in the Cretaceous period. Either they evolved from whatever it was they evolved from (and we have seen how little is known about that) extremely fast, or they hung about for sixty million years awaiting their chance. The idea that they hung about on hill-tops and descended eventually to the lowlands, where they began to prosper, has been dismissed by Professor Philip Regal of Minnesota's Museum of Natural History, who says the lowlands would have offered no particular advantage. He believes it was the rise of the birds and insects which helped to disperse their pollen more effectively as compared with the wind-dispersal upon which the gymnosperms relied.

The puzzle becomes no clearer when we look at the evolution of a single family, such as the Equidae or horses. There were many deviations from the line which was to prove the successful one. But why did *Mesohippus* (for example) diverge from the main line unless it had some advantage? Since it continued to exist for several million years, it cannot have been too badly suited to its niche. Why did it then become extinct? It is hardly likely that changed conditions accounted for this, or they would have handicapped the main line too. Did these deviants move too far in a direction which was initially favourable? It is conceivable, but no one has shown it to be so. It seems more probable that groups of genes became switched on permanently and continued to push evolution in an unsuitable direction until failure ensued. This is a radical suggestion and I shall attempt to expound it further when we come to consider the genetic aspect.

Tied up with this is the progressive and damaging increase in size which we find in creatures like the dinosaurs. This also looks like a group of genes which got switched on and couldn't be switched off, so that natural selection was impotent to regulate the trend.

When a 'radiation' occurs, when an order takes over a wide range of niches, paleontologists are given to speaking of the event as 'success'. But in what sense is this true? To remain adapted to your niche for millions of years like the ammonites without any major change implies a successful design. In contrast, to vary a design in many small ways may not be. The number of individuals of one species may be larger than the number of individuals in a whole group of species. A car manufacturer is more successful when he sells 100,000 Minis, than if he sells 5,000 each of a range of ten different designs.

The fact is, the appearance of numerous variants on a pattern delights taxonomists because it gives them plenty of work to do and enables them to accumulate files and occupy their time. An organism which just stays put is boring.

Such radiations are more probably due to the opening up of new niches, (or the vacating of niches by rivals) than to the effects of selection and variation as such. When the North American animals were able to enter South America they found many empty niches and diversified to fill them. But this doesn't really resolve the problem. How are we to account for the opossum which has changed so little that it is often called 'a living fossil' and yet which has thrown off many side-lines? If it fits its niche perfectly, such variant forms clearly would not fit, or not so well. So how could natural selection act so as to perpetuate them? I am forced to the conclusion I outlined earlier; namely, natural selection explains adaptations but not adaptation.

There can be little doubt, it seems to me, as we look back over the evidence, that the major evolutionary changes as well as the strange changes of pace in evolution, while not fatal to natural selection, necessitate postulating mechanisms that lie far outside the limits of natural selection, however valid it may be in narrower contexts.

I am very struck by the willingness of orthodox biologists to indulge in contorted and highly speculative arguments in an attempt to justify the status quo. As the University of Illinois' T. H. Frazzetta has said: 'It is incredibly alluring to think that every feature has an adaptive role . . . it is incredibly easy to rationalise an adaptive explanation for very many organic structures.' When the geocentric explanation of the solar system was breaking down, astronomers postulated more and more epicycles to explain the apparent peculiarities of the stars' and planets' movements. Finally, this was swept away and a simpler model proposed. In general, when explanations need a lot of justifying, there is something wrong with them. It is really too naive to say that, in just those periods of which the fossil record tells us nothing, the required dramatic modifications took place.

Any new explanations must lie in the field of genetics, a field which has changed out of all recognition since evolutionary theory crystallised. Not many paleontologists have kept up with it. But before we start to consider what light genetics could throw on all this, there are quite a few more anomalies in the evolutionary story to outline. They call for explanation too and at the same time may help us to see what kind of explanation is needed.

CHAPTER FIVE

Extreme Perfection

1 Inventing the Eye

Even Darwin could not quite swallow the idea that so complex a structure as the eye had evolved by the chance accumulation of favourable mutations. He called this the problem of 'organs of extreme perfection'.

The eye is not by any means the only example of an organ created by a great many changes taking place in perfect harmony, though it is perhaps the most striking. The development of the ear is equally astonishing. The formation of the circulatory system was highly complex at the chemical level. Processes of this kind present the theory of natural selection with one of its biggest challenges, so at this point I plan to examine some of them in detail and to say how the evolutionists seek to account for them, so that we can judge the plausibility of their explanations.

Eyes have, as a matter of fact, evolved on at least a dozen different occasions, though each time on a somewhat different plan. Evolutionists read this as supporting natural selection; others feel that it indicates a Lamarckian wish to achieve vision by whatever means lie to hand. Many small sea creatures, including jellyfish, have an 'eye-spot' – a light-sensitive group of cells, shielded except in one direction. This tells them the direction from which light is coming, and perhaps its intensity. Since in the sea light comes only from above, this tells them which direction is up and possibly how far away the surface is. Many planktonic forms are very precise about the depth at which they swim, and change it with the time of day.

But these are not eyes: they give no image of the surrounding world. For that you need a lens system and a large number of receptors, with a nerve-network to analyse the impulses from them.

While we think of eyes as paired, some fishes had three or four eyes, and some arthropods had six, two of them with telephoto lenses. The third eye was situated on the crown of the head and served to measure the general light level – perhaps also its direction – and was thus a descendant of the 'eye-spot'. In birds the skull is so thin that, though the third eye has vanished, light still affects the pineal gland which lies below. In modern mammals the pineal gland survives but a change of circuitry has occurred. Light detected by the ordinary eyes sends a message to the pineal, stimulating it to the behaviour which was once induced directly.

Let us start, not with the eyes of vertebrates or of insects, but with the sophisticated eyes of trilobites and shrimps, already in existence when the first crustacean fossils were formed more than half a billion years ago. It was the first use of optics in combination with sensory perception in nature and, for my money, the most incredible event in the history of evolution.

The trilobites were the first highly organised animals to populate the primordial seas and they were everywhere. The first trilobite fossils known come from the early Cambrian: they are already highly evolved so that the first trilobites must have emerged very much earlier. Their soft bodies and chitinous skins, which normally would disintegrate, have been preserved by mineralisation. (It is possible that earlier forms are missing because they were not mineralised but since so few fossils survive from the pre-Cambrian there may be some more general explanation.) The trilobites survived until the end of the Permian, a run of about 270 million years and possibly much more. They were highly successful.

They ranged in length from one-eighth to 28 inches, but most were from three-quarters to 3 inches. The eye of arthropods, I must explain, is built on a totally different plan from the human or even the reptilian eye. It consists of closely packed columns, each with its own lens at the top and its photoreceptor at the bottom, the whole protected by a cornea. They are known as ommatidia. The columns are not quite parallel but are fanned out, so that each points at a different part of the horizon. The trilobite eye was the first based on this plan. In later trilobites the eyes were in addition sometimes raised like turrets, or wrapped around the body until they nearly met, thus giving an all-round view. Specimens have been found with as many as 770 ommatidia in each eye. Eyes of this kind do not give an image as we know it – each ommatidium produces its own image. But they would be very efficient at detecting novelty, such as movement, even at a considerable distance, and would accurately report its direction.

The marvel of the trilobite eye became apparent only in 1973, when

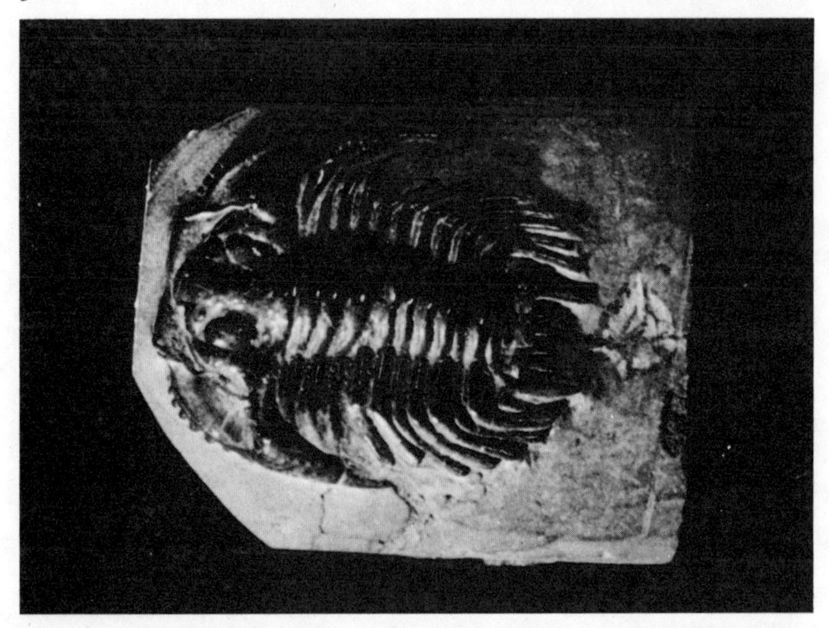

Trilobite fossil, showing raised eyes. On average, such creatures ranged in length from one to three inches

Kenneth Towe of the Smithsonian Institute reported that the lenses in the eyes of fossil trilobites consisted of precisely aligned crystals of calcite. Mounting carefully prepared fossil eyes on the microscope, he found that they produced a sharp image at distances ranging from a few millimetres to infinity, without further focusing. Up to this time paleontologists had always assumed that the calcite crystals were a relic of the mineralisation process which had preserved the whole carcass – but in that case the arrangement of crystals would have been random. Now, calcite crystals transmit light with the transparency of glass only if they are exactly aligned with the beam of light entering them. At any other angle, the light bounces off the walls and splits into various colours. Some modern arthropods have calcite crystals in their eyes, but these – and Towe finds this 'an altogether surprising thing' – are arranged randomly, and do not correspond with the ommatidia or optical units of which the compound eye is composed.

By what mechanism did these 'primitive' creatures discover how to incorporate calcite crystals, align them precisely and protect them with a cornea? Answer comes there none.

But that is only half the story of the trilobite eye. The second half likewise started in 1973 when Chicago University's Dr Riccardo Levi-Setti met Edinburgh's Dr E. N. K. Clarkson at the Oslo International Conference on Trilobites. As far back as 1901 a Swedish worker had commented on the peculiar substructure of the more evolved trilobite eyes. In 1968 Dr Clarkson began to investigate this structure with techniques, such as the scanning electron microscope, which had not hitherto been available to zoologists.

He found that the lens of these eyes was a doublet: an upper part of calcite separated by a wavy boundary from a lower half of chitin.

After Levi-Setti had delivered his lecture on the amazing light-collecting properties of the trilobite eye, he had coffee with Clarkson, who told him of his work. Levi-Setti at once had a hunch that this doublet structure must represent some form of optical correction. Armed with sketches made on the paper napkins of the canteen, he returned to Chicago, where he brooded over the problem. One day, reading that bible of all students of vision, the *Traité de la Lumière* of Christian Huyghens, published in 1690, he found the description of an aplanatic or spherically corrected lens 'which resembled unmistakably the wavy shape seen in Clarkson's sketches'.

Huyghens' mention of some earlier results by Descartes led him to peruse the latter's *La Géométrie*, published in 1637. 'There I found a second construction somewhat different from that of Huyghens, but designed to perform the same function. This matched a second version of the trilobite lens-shapes described by Clarkson. Armed with the conviction that trilobites had solved a very elegant physical problem and apparently knew about Fermat's principle, Abbé's sine law, Snell's laws of refraction, and the optics of birefringent crystals, I set out to inform Dr Clarkson of the meaning of his trilobites' lens-shapes.'

Thus the trilobites evolved a lens shaped to correct for optical aberration identical to that proposed (quite independently of any knowledge of trilobites) by Descartes and Huyghens half a billion years later.

Why was such perfection needed? Dr Clarkson suggests that trilobites may have lived in very muddy, turbid water. Or perhaps they only came out at night or at dusk. The thick lenses, thanks to the optical correction, would be more efficient light-collectors. But to make the matter more puzzling still there is the fact that some trilobites were blind.

How did the earliest trilobites collect together the intricate genetic information needed to construct this semi-miraculous structure? And how strange that all that know-how should have been lost again when the phacopid line of trilobites became extinct at the end of the Devonian. Its collection argues for the existence of some directive force, but its abandonment argues against the existence of a coherent plan.

One could perhaps accept the idea that a happy accident caused the sensitive spots to become more numerous, and that another happy accident placed them at the bottom of tubes. It is a little harder to accept that by chance the tubes were not parallel, which would have been the obvious pattern of repetition, but were slightly divergent. But by what conceivable chance could the trilobite have accumulated the one material in the universe – namely calcite – which had the required optical properties and then imposed on it the one type of curved surface which would achieve the required result? There are innumerable possible shapes, none of which offer the unique advantage of spherical correction, except the one I have described.

Now if anything were needed to cap this evolutionary feat it is the fact that the refractive index of calcite and of chitin are precisely those needed to produce an aplanatic lens.

Probably about the same time as the trilobites were achieving their breakthrough the shrimps were attempting the same thing, but they used a form of mirror optics, the mathematics of which were only worked out in 1956 as a consequence of interest in optical fibres. Twenty years passed before Michael Land of Sussex University showed that instead of using transparent lenses to refract the light, the shrimps bounced it off mirrors set radially. Since shrimps don't secrete metal, they relied on multi-layers of guanine and water – the same technique makes fish scales silvery. The realisation that they had used a technique never attempted by man only dawned in 1975; it showed they could not be closely related to the ephausids or krill with which they are classified and raised new problems for the theory of evolution by natural selection.

While we are still reeling at the improbability of this, let us pass on to the much more complex mammalian eye, such as our own. Its basic design first appeared in the fishes but since then it has been improved from time to time.

2 *Re-inventing the Eye*

You are probably familiar with the general structure of the human eye, which has so often been described: a ball with a lens on one side and a light-sensitive retina on the other, the retina being made up of two kinds of light receptor, known as rods and cones. A transparent cover, the cornea, protects the lens from damage, and an iris diaphragm – very like those used in many cameras – to exclude excessive light which would simply bleach the light-sensitive pigments in the retina. There are muscles which change the shape of the lens to preserve focus, depending on whether remote or nearby objects are being scrutinised, and other muscles to move the eyeball. The ball is filled with an aqueous humour which is replaced every four hours. There are tear glands to wash the eye, and an eyelid fringed with eyelashes to protect it from dust.

The saltness of our tears reminds us of our evolutionary origins. For the eye originated in salt water: when the fishes moved on to land, they had to surround it with its original medium to prevent it drying out.

The essential features of the vertebrate eye appeared quite suddenly in evolution, so much so that a German anatomist was led to compare the event with the sudden appearance of Athena, fully armed, from Jove's forehead in Greek myth. And while we can perhaps find precursors for the photosensory cells in the eye-spot of *Amphioxus* or those of the jellyfish, there are no precursors for the lens, the origin of which, in the words of Gordon Walls of Wayne University, who has made the study of the vertebrate eye in all its forms his life work, is 'a tantalizing mystery'. (Part of the mystery is how the lens comes to lie inside the coats or 'tunics' of the eye, which derive from the meningeal coats of the brain. It is, so to say, a bit of skin which has got inside the coating of the brain.)

In the course of evolution various refinements were added, notably the ability to distinguish colours. Less well known is the fact that the eye employs various red and yellow filters to enchance acuity. The pigeon, for instance, has yellow filters which take out the blue of the sky in that part of the retina it uses when looking upwards, and red filters which separate the greens in that part it uses in looking downwards.

Originally these filters took the form of oil-droplets within each photosensitive rod or cone. When these were lost, a yellowed cornea was evolved to achieve the same end. Burrowing species often protect their eyes with 'spectacles' which cover the cornea and these can be yellow-tinged too.

Other striking examples of adaptation are the devices which help nocturnal animals to see in the dark, and those which protect tropical, diurnal animals from too-bright light. Nocturnal animals, as you might expect, have a preponderance of rods which makes for greater sensitivity at the cost of unsharp vision; this is why nocturnal animals depend largely on their noses and do not try to eat small objects like seeds. The original chordates were bright-light animals, equipped with cones. The fishes invented rods in order to see something in the murky ocean depths and perhaps to lengthen their effective day.

Great sensitivity at night also necessitates devices to protect the

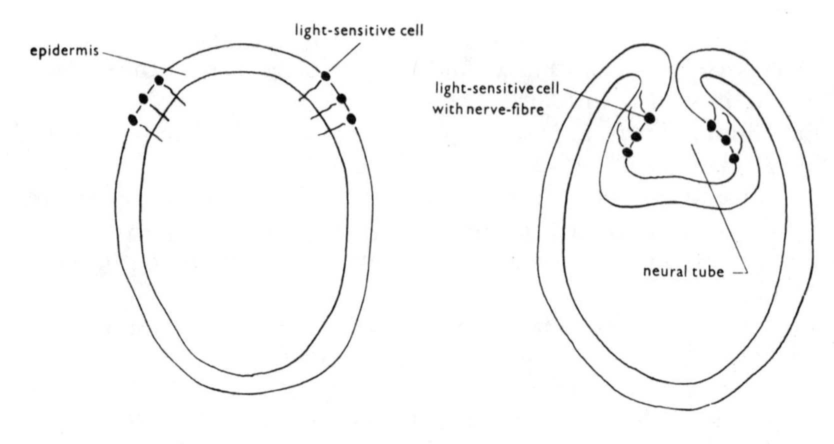

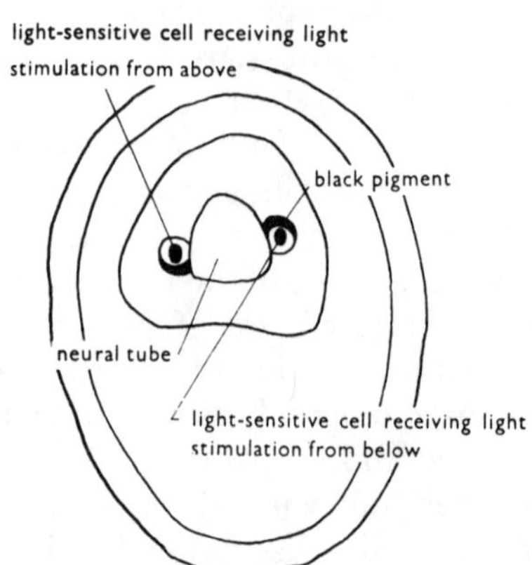

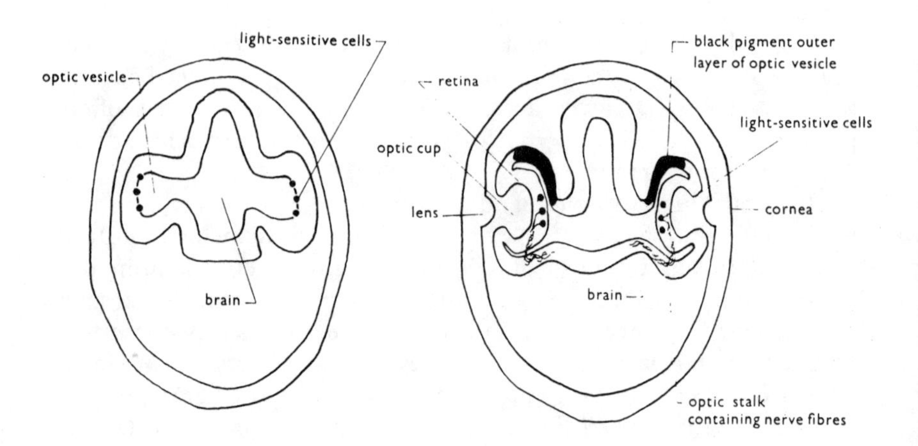

Changes in the structure of the eye from that of a primitive vertebrate, showing the folding of the neural tube to form the mammalian brain and eye

retina by day. Among these was the slit pupil which can close off light entirely, unlike the iris diaphragm with which we humans are equipped. Another device for increasing efficiency in the dark is the tapetum, a mirror behind the retina which reflects back any light which has passed through it and thus gives it a second chance of detection. Anyone who has seen a cat's eyes reflect the light of motor headlights at night will know what I mean. Unbelievably intricate forms of tapetum are found in some fishes. Silvery crystals of guanine serve as a backing for the visual cells. But granules of the pigment fuscin emerge to cover them up in bright light, and retreat to expose them in poor light.

Raptorial birds like the falcon and eagle, as you will know, have acute vision – as much as eight times the resolving power of human vision. This is achieved primarily by an intense concentration of receptors, as many as a million per square millimetre. And of course these receptors are mainly cones.

One of the great puzzles in the evolution of the eye is how the eye of snakes arose, for their visual cells have no similarity to those of the lizards from which snakes derived, and this is true even of the most primitive forms. Gordon Walls calls it 'literally a bundle of substitutes for lizard-eye features'. Snakes arose from huge monitor lizards which learned to live in burrows in the dark. Their eyes degenerated: there was no need for an iris diaphragm or focusing muscles. The visual cells themselves lost pigment. There was no need for a lacrimal gland. When the burrowing lizards re-emerged as pythons and boa constrictors 'the snakes had almost to invent the vertebrate eye all over again. Nothing like this tremendous feat has occurred in any other vertebrate group, as far as we can tell,' says Walls. And they did so in a slightly different manner from before.

It is just about possible to accept the idea that all these beautiful adaptations – and I have mentioned only the most obvious – were brought about by natural selection. It is harder, I think, to see how major structures like the eyelid appeared, for a small flap of skin would offer little advantage. Until an eyelid covers the eye, it is not much use. But what is really hard to swallow is the structure of the eye itself.

As far as I know, no one has estimated the number of mutations which would be necessary to bring about all these changes, and not only changes but the creation of new structures (such as the iris) for which there was no precedent. Yet the essential features of the eye appeared quite abruptly in evolutionary terms. As we shall see later, this is not really long enough, bearing in mind that many mutations would do more harm than good. Moreover, the more complex the structure gets, the more precisely engineered must be any addition to it. In the end, it is a question of chance, and all depends on whether you think such an amazing sequence of lucky turns is credible or not.

Some authorities believe that we can learn more about how the eye first appeared by studying how it is formed in embryonic development. Even before the brain is formed, two dimples can be seen in the nerve tissue which are destined to become the retina. As the brain forms, two bulges appear, become cup-like and form the retina. They secrete a substance, termed an inducer, which causes the skin above them to develop into a lens. If an eye-cup is transplanted to some other part of the body, it causes a lens to form there. This accounts for the way in which brain tissue and skin cooperate. But how in Darwin's name did evolution produce a substance with such extraordinary properties? Its mode of action remains a mystery.

Evolutionists believe that they can account for the formation of the eye, nevertheless. They can assemble eyes from living creatures of ascending orders of complexity and believe that these indicate what may have happened over evolutionary time, starting with a single pigmented cell or group of cells in a creature which is already transparent and proceeding by gradual changes to the elaborate eye of mammals.

The weakest feature of this hypothesis – and it is only a hypothesis – is that it demands a long time for the eye to become reasonably efficient whereas the eye appears quite suddenly in evolution. The earliest fishes have quite sophisticated eyes. Also there is the fact that the vertebrate eye is paired, unlike any of the light-sensitive organs which preceded it.

In the lancelet, the forerunner of the fishes, there is a light-sensitive organ – you could hardly call it an eye – consisting of a large cell with a pigment cell draped over it, so that the latter receives light from only one side – thus giving the lancelet an idea where the light is coming from. Some of these cells face up, some down. Its body is transparent. The next stage is found in ascidians, which have no eyes as adult but have something like an eye in the tadpole stage: a cup-shaped layer of light-sensitive cells topped by three transparent cells which might be called a lens *in posse*. Then suddenly in the lampreys there appear not one but two pairs of eyes: two ordinary eyes and two pineal eyes.

Sir Gavin de Beer, now dead but for long Director of London's Natural History Museum, defended the orthodox view by assembling in his *Atlas of Evolution* a series of pictures showing a sequence starting with the primitive eye-spot and culminating in the mammalian eye. This is disingenuous, for, as Walls says, it is useless to trace the vertebrate eye back to the pre-chordates. De Beer then comments. 'There can be little doubt that the series of stages just described through which the eye passes in embryonic development is a repetition of the manner in which it evolved.' But elsewhere he subscribes to the well-known fact that ontogeny does not recapitulate phylogeny in an exact way. (For instance, in the embryo teeth develop late in the story, whereas in evolution they appear early.) In any case, to show the stages through which development passed does not explain *why* it took this course or how the jumps to new levels of sophistication were achieved. The problem cannot be brushed under the carpet in this way.

Having disposed of the eye, writers like de Beer appear to think that they have disposed of the whole problem of coordinated development. But of course almost all evolutionary changes involve the coordination of many different changes. If the leg grows longer, both bone and muscle must lengthen, not to mention the venous system and the nerves.

Precisely at the time the eye was developing, so was the ear and other sense organs, such as taste. The fishes enter on the scene equipped with an array of sense organs. We have to believe that mutations bringing about all these changes, as well as the structural ones, occurred simultaneously. The development of the ear is hardly less complicated than the development of the eye.

If the eye is the most complex instance of coordinated development, the ear is perhaps the most puzzling, because of the extensive way in which existing structures, originally intended for another purpose, were remodelled to detect and analyse sound.

3 *Mobilising the Ear*

Bear in mind that the ear has a double function: it is both an organ for detecting sound and one for preserving balance by analysing the movements the body is undergoing. Basically, it analyses sound frequencies by a tubular organ containing hairs ranging in size from large to small. Each hair resonates to a different frequency, i.e. to a different pitch. In mammals this tube is rolled up into a spiral known as the cochlea, from its resemblance to a snail. The cochlea of mammals contains also the organ of Corti, of which more anon. At the same time the ear analyses accelerations by means of three semicircular canals. Sensitive hairs detect the motion of fluid within them as the body moves and the brain integrates the three sets of signals to arrive at the vector. Finally, within a cavity known as the lagena, a small lump of stone or mineral concretion rests on sensitive hairs and tells the brain which way is up.

I shall not touch on the various membranes and canals which

subserve these structures, just as I shall not describe the cells of Deiters and Henti's stripe, for my purpose is not simply to impress you with the complex refinement of these mechanisms but merely to ask where on earth all this hardware came from. How was this complex set of structures evolved?

To find an answer to that question we must go back to the earliest bony fishes. They possessed, as do modern fishes, a structure known as the lateral line, running from head to tail on either side. At the head end it fans out into a system of canals in the skull. The lateral line is a canal enclosing cells armed with sensitive hairs embedded in gelatine which detect the motion of the fluid in it and which in this way register vibrations in the water or so it is claimed. It has also been alleged to register electric currents or to be an organ of taste, registering chemical changes. But since recent experiments have shown that it can distinguish between water flowing past it head-to-tail from water flowing tail-to-head, the most probable explanation is that it is an accelerometer for straight-line swimming.

Where it came from is totally mysterious, but where it went to is the ear. The transformation took place in six or seven stages. As evolution proceeded, the canals in the head migrated into the bony structure of the head and bent round to form semi-circular canals. The ciliated cells were now enclosed in a capsule, still with their ends in gelatine. Other parts of the lateral line formed into closed sacs: the sacculus, the utriculus and the lagena. In bony fishes, the utricle was and is the main detector of posture, for one of these sacs became loaded with a mineral concentration or crystal which, resting on sensitive hairs, indicated accelerations and decelerations and also which way was up.

When we turn from fishes to terrestrial animals we find the sacculus and utriculus have been carried over unchanged, but the lagena has wound itself into a spiral. This first occurs in snakes, the process being completed in birds and mammals, where the spiral becomes a helix: in a word the cochlea aforementioned. In mammals the sensitivity of the cochlea is improved by a new structure, the organ of Corti, which renders the hairs so sensitive that they can detect vibrations whose amplitude is no more than the diameter of a hydrogen atom. What was once merely a pressure detector now provides for us the miracle of sound.

So much for the inner ear. The eardrum first appears in the frog, lying flush with the skin: snakes have no ear-drum. Then various jaw bones are adapted to transmit the vibrations of the eardrum to the inner ear. Meanwhile the efficiency of the cochlea is improved by the appearance of the basilar membrane.

Finally the external ear is added to improve directional hearing.

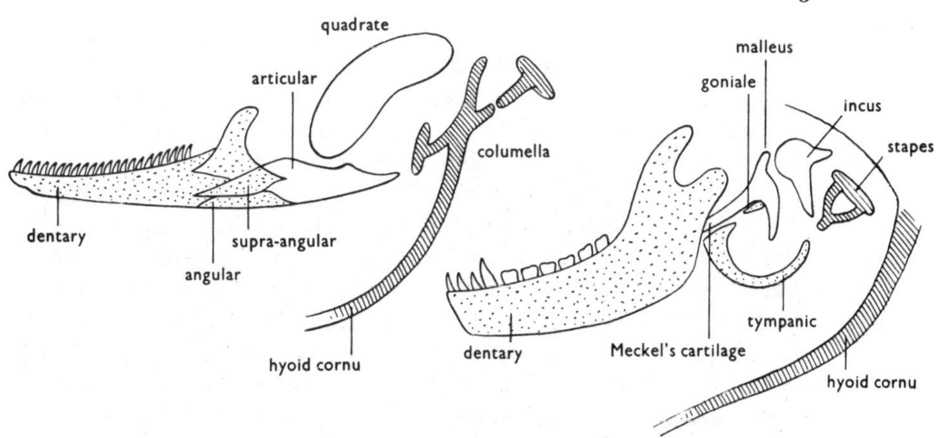

Comparison of the reptile (left) and mammal jaw. The bones which form the hinge of the reptile jaw become incorporated in the chain of ear-ossicles in mammals. Thus the mammal's goniale, tympanic, stapes, incus and malleus are equivalent to the reptilian supra-angular, angular, columella, quadrate and articular respectively.

With all this, of course, went improvements in the brain, most notably the power to compare the times at which signals from one source reach each ear, thus providing a method of estimating the direction in which the source lies.

Thus, in the course of evolution, there were six major developments, two of which occurred in the fishes, two in the amphibia and two in mammals. Such, at least, is the account given by people like Willem van Bergeijk, of Bell Telephone Laboratories, who is the acknowledged authority. But the eminent morphologist J. W. Torrey is not convinced. 'The evolutionary origin of the inner ear is entirely unknown,' he insists.

In contrast with the case of the eye, where undifferentiated cells were specialised into the required forms, here existing structures have been profoundly modified and even shifted to another position in a progressive series of changes which certainly look more like the refinement of a plan than the result of a series of happy accidents.

But the insoluble problem is how and why did a balance organ become an organ of hearing? As van Bergeijk pointedly asks: 'What prompts the fish to begin developing a sensory apparatus that will respond to a stimulus about the very existence of which the fish knows nothing?'

Van Bergeijk believes that the original balance organ would never have evolved mechanisms for hearing but for the emergence of the swim bladder. The original purpose of this organ is to enable the fish to adjust its density to the density of the ambient water and so control the depth at which it swims. Since the bladder is sensitive to changes in external pressure, it vibrates in harmony with pressure changes in the water. In time these vibrations came to excite the ear. Hearing as distinct from the mere detection of pressure waves, was born.

After describing the last part of this process, the adaptation of the bones linking the jaw to the skull into a chain of ossicles linking the eardrum to the inner ear, Ernst Mayr sweepingly remarks: 'Not all the steps in this process are yet entirely apparent, but I think little doubt is left as to the principle involved.' If by 'principle' one means merely progressive remodelling, the statement is a truism. But if 'principle' means that chance selection brought about these elaborate changes, then there must be very great doubt indeed. Like de Beer, Mayr does not seem to appreciate the elementary point that demonstrating the occurrence of a sequence of events does not explain why they happened.

But what kind of mutations could bring about the major changes I have described? Could cause a tube to roll up into a helix? Could cause other tubes to form semi-circular canals accurately set at right angles to each other. Could grade sensory hairs according to length? Could cause the convenient deposit of a crystal in the one place it will register gravity? Even more amazingly, some fishes do not trouble to secrete a crystal but incorporate a bit of sand or stone. What kind of mutation could achieve this – when and only when a natural crystal is not formed? The purpose is fulfilled, the means are unimportant. It just doesn't make sense.

The eye and ear are outstanding instances of complexity, and I have already commented elsewhere on the complexity of the readjustments attending so relatively straightforward an adaptation as the lengthening of the legs and neck of the giraffe. But before we abandon the subject I would like to present you with two more instances of a rather different kind, namely the reorganisation of chemical processes which is sometimes involved in an evolutionary change and the strange phenomenon of mimicry.

4 *Blood and Iron*

While structures such as the eye and ear arouse our wonder because we are familiar with their appearance and also with what they

Opposite Model of haemoglobin molecule seen (above) from top and (below) from side in diagrammatic form. Represented in each are two alpha chains and two (darker) beta chains, each enclosing a disk depicting the molecule's container of iron

N
C
C
N

achieve, there are other examples of subtle coordination which are less obvious but just as remarkable. The formation of blood, for instance, is a saga in itself. Starting as sea water, it evolved to a complex liquid containing at least eighty components, many of them still insufficiently understood. A component of central importance, of course, is the haemoglobin which picks up oxygen in the lungs, while giving up carbon dioxide; and then having travelled to the muscles, gives up oxygen and accepts carbon dioxide, which the muscles produce as a result of burning fuel, much as a car produces carbon monoxide. It is a remarkable molecule indeed which at one moment has an affinity for oxygen and a few seconds later loses that affinity; that it simultaneously changes its preferences with respect to carbon dioxide make it even more remarkable. There could be no more amazing example of adaptation to a task. Blood carries fifty times as much oxygen as sea water.

But even this is only part of the story. The core of the haemoglobin molecule is an atom of iron. How does it get there? The human body contains about 4.5 grams of iron – less than half the permitted basic weight of a letter sent airmail. The iron we eat – it is present in some vegetables – is in the form known as ferric and must be converted to the ferrous form before it can be absorbed. But oxygen tends to reverse this, driving ferrous iron to ferric. To prevent this, the intestine had to be made acid. Cells evolved which secreted hydrochloric acid, which had this effect. They appear first in the chordates. The mucosal cells then adapted so as to absorb iron. But iron tends to add on to proteins and would never reach the liver where it is stored, if it were not for a substance named siderophilin, which appeared somewhere along the line. Then, as was discovered in 1937, iron is stored in the form of ferritin, a substance acceptable to the body which incorporates iron freely.

Haemoglobin makes a haphazard appearance in the evolutionary story, appearing in many different phyla. It is found in some species of *paramecium* (very primitive single-celled creatures known to every school biology class). It is found in worms, molluscs, insects and even in the roots of leguminous vegetables. What it is doing in all these places is largely unexplained. One thing seems clear, that it was invented time and time again, quite independently.

Surely it is easier to suppose that the genetic system contained a group of genes capable of eliciting the production of haemoglobin when the appropriate stimulus was applied, than to suppose that the same group of mutations occurred by pure chance again and again.

Vertebrates, however, needed a better blood supply for their larger brain (or so we may suppose) and this explains, perhaps, why they made the most use of it.

5 Mimicry

A young English naturalist named Walter Bates spent the years 1849 to 1860 forcing his way through the forests of Brazil in search of butterflies. In 1862, when he was thirty-seven, he published in the *Transactions* of the Linnean Society of London an article which for more than a century has provoked hot discussion, not only among scientists but among philosophers and theologians, to say nothing of amateur naturalists.

He captured thousands of butterflies, which he classified in ninety-four species, the whole forming a family he called Helioconides. But to his amazement, some of the butterflies which looked identical proved on close examination to be constructed so differently that it was clear they must belong to different families. The explanation he put forward was this. The Helioconidae fly quite slowly, are brightly coloured and therefore easy to catch. Birds, nevertheless, learn to avoid them because they have a most unpleasant taste. By evolutionary chance some butterflies in the family Pieridae came to resemble the Helioconidae in appearance and hence birds began to avoid them, though they taste delicious. (As a matter of fact, tasting new butterflies is a regular activity among butterfly-hunters; our own Professor E. B. Ford was an enthusiastic butterfly-cruncher.)

This phenomenon, which today is known as Batesian mimicry – for as it turned out there are other ploys in the mimicry field besides imitating unpalatable models – aroused widespread incredulity, and nearly half a century after Bates' bombshell of an article, it was still being attacked on the grounds that there are many examples of mimicry in nature which appear to be pointless as far as survival goes. Thus some marine algae look like ferns. There are ant mimics which have no predators. But against a few such chance resemblances, naturalists were able to put a growing list of resemblances which seemed genuinely advantageous.

The range of such imitations is startling. There are insects which look like sticks, and others which look like leaves; caterpillars which resemble sprouting twigs, and bugs which look like thorns; moths which look like bird-droppings, and frogs which look like leaves. There are even fish which imitate dead leaves. Cleverest of all, perhaps, is an insect pupa which looks like a snapped-off twig. Compared with these, the striping of the tiger, and even the colour changes of the chameleon or some bottom-lying fishes seem amateurish.

Sometimes the resemblance is auditory rather than visual. Some hole-nesting birds, such as chickadees and titmice, hiss like snakes when disturbed. Frequently the predator is deceived by behaviour.

Sea dragon; an outstanding example of camouflage based on outline

An unattractive beetle named *Eliodes* stands on its head when threatened, the better to squirt a repellent from its abdomen at the intruder. Another beetle, *Megasides*, stands on its head, in imitation, although it has no spray-gun, hoping the bird will be fooled.

Such changes certainly look purposeful but are explained as being the result of natural selection. The difficulty is that a whole suite of changes is necessary. The leaf insect has got to be flat, oval, green and have surface markings somewhat like the veins in a leaf, to say nothing of having (as some do) a stalk. Any one of these changes alone would make it more conspicuous, not less. Moreover, it has to learn to stay still. A leaf which hopped across open ground and plunged into a pond, or one which took off and flew in a determined manner, would attract attention to itself from a human being, let alone a predator.

There are other problems. Birds which, like cuckoos, place their eggs in other birds' nests, begin to produce eggs which resemble quite closely those of their host in colour and markings. As there is no evidence that host birds turf the foreign eggs out of the nest it is hard to see where the selective advantage of such colouration lies, quite apart from the delicacy of the physiological adjustments required.

It is not clear that natural selection requires the mimicry to be as perfect as it is. Several lepidopterists have tried to measure how far the similarity actually benefits the butterfly. For instance Jane Brouwer and her husband caught butterflies in the West Indies and froze them with wings outstretched. They collected good mimics and bad mimics, as well as a brown butterfly as a control, and offered them in turn to a bird next morning, the bird being in a large cage. Several birds were used with due precautions. They found that even a moderate degree of similarity offered some protection, from which they concluded that a micro-mutation in the required direction would offer some selective advantage and so start the ball rolling. But if birds are not very discriminating, why is it necessary for the imitation to be pushed to such an extreme?

And while such laboratory experiments seem to confirm the working of natural selection, it is curious that the non-mimicking varieties seem to survive just as well in real-life conditions. Moreover, a disagreeable taste seems to offer much better protection. So why not evolve it, as other butterflies have, instead of doing the much more elaborate trick of imitating them?

While mimicry is commonly used to escape predation, it has other uses. One shrimp-like creature resembles its intended victim, and gets close to it without being spotted: a marine wolf in sheep's clothing. Jennifer Purcell of the University of California, scuba diving off California and Massachusetts, has examined the diet of siphonophores, a creature not unlike a jellyfish. (The Portuguese man-o'-war is the most widely known example.) Their stomachs are equipped with batteries of the stinging nematocysts we met in Chapter One. But the stinging apparatus curiously resembles the

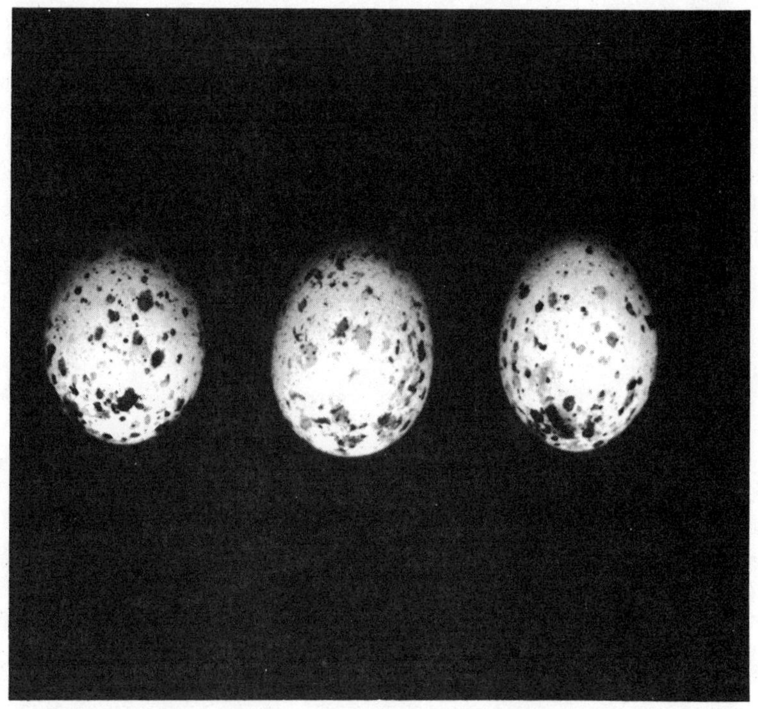

A cuckoo's egg (centre), adapted in this case to resemble the eggs of a reed warbler

copepods and larvae which are common in these waters; it even has two dark spots resembling eyes and filaments like antennae. The lure appears to dash about as the tentacles stretch and contract. The fish which seek to eat them, themselves fall prey to the siphonophores.

But the delicacy of the mimicry process goes much further. For instance, in regions which are snow covered in winter but not in summer some birds and mammals assume a white or whitish colouration in winter but change back to brown in summer. The arctic fox and the ptarmigan will serve as examples. Here we have the much harder problem of explaining how the change becomes regulated. I can imagine a gene for becoming brown, but hardly a gene for becoming brown in summer only. The effect may be temperature dependent. But how did it evolve? A ptarmigan which simply mutated to white would be killed off in the summer. Unless the temperature mechanism evolved simultaneously and in the same individual there would be no evolutionary advantage. And it is odd that the male ptarmigan changes colour later than the female, remaining white some weeks longer than it is wise to do so. This is 'explained' by saying that it is more important for the female, as the egg-layer, to survive. But males have their uses too. Rather a thin explanation, I fear.

Another problem: why do not other species make similar arrangements? Grouse, for instance? And, if protective colouration is so advantageous, why do not ptarmigan become commoner than grouse instead of being rarer?

Polar bears are white, of course. Or rather, they are a yellowish-white, more like thick ice than snow. But there are no creatures in the Arctic powerful enough to prey on polar bears, so why have they adopted protective colouration? Surely it was not just to be in the fashion? Pushed into a corner, the natural selectionists suggest, rather frantically, that it deceives the seals when they come up to breathe at a hole in the ice, enabling the bear to scoop them out with one back-breaking blow.

It is not difficult to accept that some simple colour changes are advantageous. For instance, snails which live in woodlands tend to have darker shells than their cousins which live in the open. Experiment shows that thrushes attack light-coloured snails in woodlands more often than dark-shelled ones. And conversely in the open. And I have already described the melanism of the moth *Biston betularia*. Here only a single character, possibly controlled by a single gene, is in question. The leaf insect, with its suite of changes, is a rather different matter.

In fact, why is not protective camouflage far more widespread generally? It would have paid the caveman, when stalked by the sabre-toothed tiger, to resemble a boulder or a tree. The falcon might ignore the hare if it looked like a piece of wood. (A trick some alligators have learned, though they are not much subject to predation.)

In short, even if evolutionists have demonstrated the advantages of camouflage they are very far from having demonstrated that natural selection brought it about.

These are superb examples of coordinated development but almost every evolutionary change calls for modifications of several structures, and often of physiological processes, more or less simultaneously. To give the bird's beak the power to crack nuts and hard seeds meant the fusing of its hinge bones to the back of the skull as well as changes in the beak itself. In his *Complex Adaptations in Evolving Populations*, T. H. Frazzetta of the University of Illinois has analysed many such adjustments, saying 'There is a definite suggestion that certain modifications cannot greatly precede or lag behind others but must keep pace if the performance of the machine is not to become sloppy.' As he aptly says, the task is to improve a machine while it keeps running.

There is even an oyster-like creature equipped with a hinge. The upper section evolved two projections and these fit loosely into projections on the lower section. As Professor Grassé remarks: 'It is hard to see how natural selection could have presided unaided over

Two examples of the peppered moth, *Biston betularia*. Both light and dark specimens have evolved in recent years in response to changes in their surroundings

such delicate fittings as the double lock of the ant-lion's mouth, the hinges of the bee's wings, the push-button of Nepas, which have to be perfect if they are to function at all.'

Many of these modifications possess an all-or-nothing character, which makes it very difficult to understand how natural selection could have produced them. Thus the python's upper jaw is hinged, and this affords an advantage. The python possesses two curved, pointed fangs which sink into the prey and hold it while the python begins to swallow it. However, the teeth are set pointing backwards, so that they would never enter the victims' flesh, were it not for the hinge, which causes them to strike at a more obtuse angle than they otherwise would. This hinge was not of the slightest value until it formed and worked. How could it have appeared by 'infinitely small gradations'?

CHAPTER SIX

Puzzles and Plans

1 Straight-Line Evolution

Over the years, a significant number of biologists have found it impossible to believe that the marvellously precise adaptations to be found throughout the world of plants and animals could be due solely to chance and have come, often reluctantly, to the conclusion that there exists some kind of plan or purposiveness in evolution. I have already mentioned some of their names and others will appear in this chapter.

Unfortunately biologists tend to confine themselves to their speciality, so that the facts which convinced them are dispersed in many scientific papers. I try here to bring some of them together for assessment.

Looking over the mass of material I perceive four fairly distinct sets of puzzling facts: (1) The existence of trends which continue for millions of years – for if evolution is a random process one would expect purely random changes of direction. (2) The repeated occurrence of the same evolutionary process; or, the simultaneous occurrence of the same process in widely divergent species. (3) The appearance of structures before the need for them arises – this is the most convincing evidence of all. (4) Adaptations and developments, such as the development of the eye, which call for coordinated changes which could hardly occur by chance. These last we have already discussed.

Let us start by looking at the idea of evolution in a straight line, or orthogenesis.

Around 1895 the American paleontologist Henry Fairfield Osborn, who died in 1935, became convinced that evolution was 'a creative process . . . proceeding in the direction of future adaptation'. He christened the process orthogenesis and wrote many books and papers about it. Evolution, he believed, led to the production of better and better forms, a happy circumstance which he called 'Aristogenesis'. As he was supremely eminent, with degrees at Cambridge as well as Princeton and Yale; as he was president of the American Society of Naturalists, the American Society of Paleontologists and the American Association for the Advancement of Science, a winner of the Darwin medal and a Fellow of the Royal Society and goodness knows what else, his views were listened to with attention.

Paleontologists tended to accept 'straight-line evolution', since they were aware of the evidence for long-continued trends; Darwinians were rootedly against it. California's Verne Grant declared: 'There is not a shred of evidence for orthogenesis in any of its forms.' Being a paleontologist, G. G. Simpson is much less dogmatic. There is 'no possible doubt', he says, 'that some degree of rectilinearity is common' – especially at family and generic level, though he is quick to emphasise his orthodoxy by adding that the evidence conclusively opposes the presence of 'a primary directing principle'.

The trouble is that orthogenesis could mean so many things. There are certainly many long-continued trends in evolution. We have seen how the development of the eye and ear proceeded over millions of years. Indeed, evolution itself is a long-continued trend. But these trends were not especially uniform; the eye took many forms and sometimes degenerated or vanished. Frequently cited as an instance of orthogenesis is the development of the horse family, from *Eohippus* to the modern form, an example chosen mainly because of the unusually complete series of fossil skeletons available. There were many branches from the line which was to prove the successful one. This certainly does not suggest the existence of a plan dictating the horse's evolution.

On the other hand, the horse's teeth became progressively better adapted to grazing (starting from a stage in which they were adapted to browsing on shrubs or trees). Darwinians point to this as an example of natural selection at work, and perhaps it is. But natural selection postulates the occurrence of a mutation making the horse's teeth more suitable to grazing at one jump, rather than a gradual modification over millions of years. Are we to understand that a whole series of mutations took place, each edging the tooth nearer the ideal form? Though it is true that more than one mutation occurred – for instance, a new sort of dental cement suddenly crops up in *Parahippus* – it is not easy to account for these slow trends.

This is even truer of the evolution of the horse's hoof. The gradual retreat of the two side toes and modification of the middle toe into a

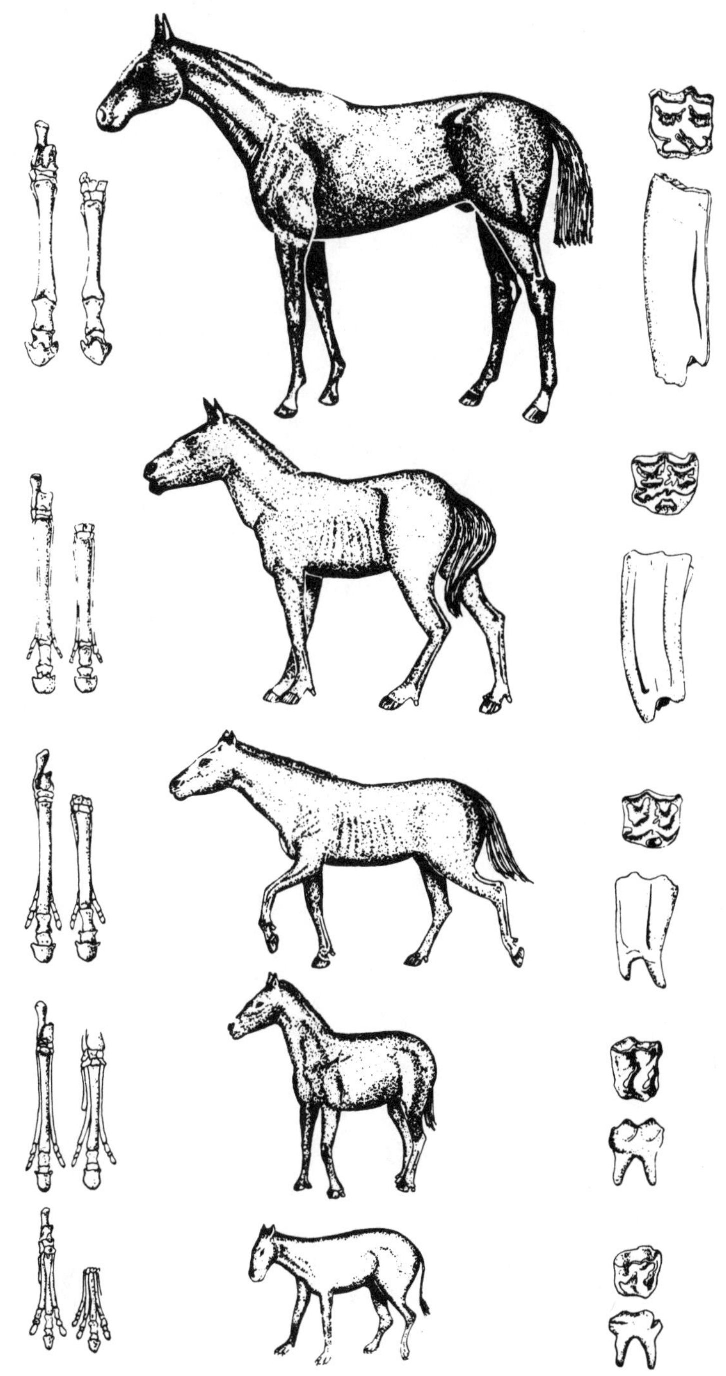

Steps in the evolution of the horse and its modern relatives, showing in each case the hind and fore foot and, from above and one side, a molar. In ascending order: *Eohippus*, *Mesohippus*, *Merychippus*, *Pliohippus*, and *Equus*

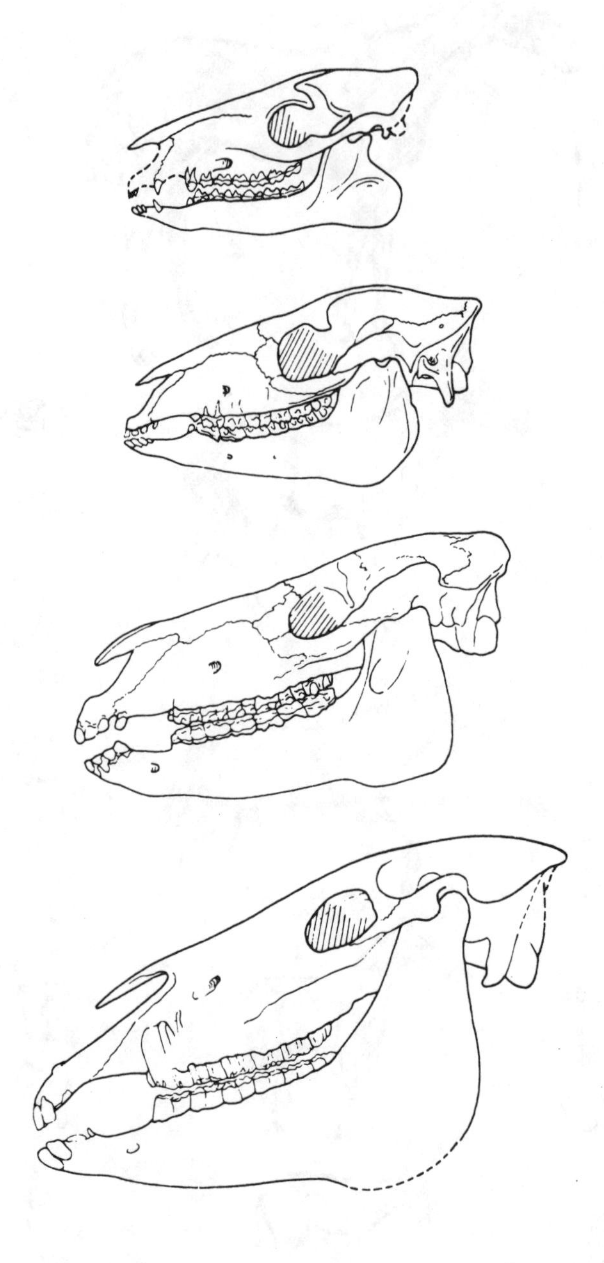

Examples of the evolution of dentition among ancestors of the horse. From top to bottom: *Eohippus*, *Mesohippus*, *Parahippus*, *Pliohippus*

hoof provides us with a classroom demonstration, as vivid as it is rare, of evolution at work. Whatever the explanation, it could certainly be described as a trend, even an orthogenetic one.

In short, even if we reject the mystical claims of a Teilhard de Chardin, or the almost equally insubstantial claims of Henry Fairfield Osborn, there remains a process to be reconciled with natural selection or to be explained in some other way.

The most obvious trend, and the one of which there are most instances, is the trend that is so often found towards larger and larger size. The horse of today is much larger than the dog-sized *Eohippus*. True, some of the sideshoots were even smaller than *Eohippus*, but the successful line tended towards greater size. Such trends are odd enough to deserve a little discussion. And here we might digress for a moment to Henry Cope and his law of body size.

There have not been many swashbuckling characters in the history of paleontology, but Henry Drinker Cope was one all right. Another was his arch-rival Othniel Marsh. Before he took up paleontology, Cope had won a reputation as the greatest comparative anatomist in the US. The Middle West of the US was, all unsuspected, a paleontological paradise. The bed of a dried-up sea, it was littered with the bones of gigantic animals, still unknown to man: *Stegosaurus*, *Brontosaurus*, *Diplodocus*, *Triceratops*. In 1868 the thirty-eight-year-old Othniel Marsh found a small hollow fossil bone, in a dried-up river bed, which looked like the tibia of a bird. But no living bird possessed a joint such as this bone displayed, permitting a freedom of movement in one direction 'that no well-constructed bird could use on land or water'. It came, Marsh showed, from a toothed pterodactyl of fish-eating habits, with a wing-span of twenty feet. He named it '*Pteranodon*'. Thomas Huxley commented: 'The discovery of the toothed birds by Mr Marsh removed Mr Darwin's proposition that many animal forms have been utterly lost from the region of hypothesis to demonstrable fact'. Darwin wrote to him: 'Your old birds have offered the best support to the theory of evolution.'

Marsh also found bones which could only have belonged to a primitive horse, standing three feet high. He called it *Protohippus*. During the next six years a series of specimens in a sequence from *Protohippos* to the modern horse was discovered. In this sequence there seemed to be a long-term trend moving inexorably towards the horse as we know it. The feet, originally three-toed, became one-toed and were equipped with a hoof. The teeth were adapted for grazing, rather than browsing. The skull lengthened into the shape we know so well. This sequence was to prove exhibit No. 1 in the case to establish 'orthogenesis' or evolution in a straight line – something

A reconstruction of *Brontosaurus*, a creature some sixty-five feet in length

Reconstruction of *Stegosaurus*, a dinosaur of the Jurassic measuring eighteen to thirty feet long

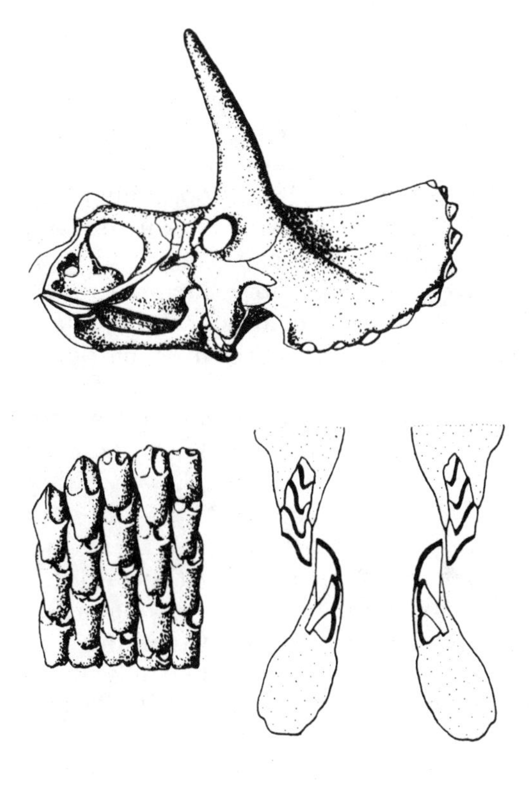

Skull of *Triceratops*, with part of its battery of replacement teeth, and cross-section showing the relation of upper and lower jaws

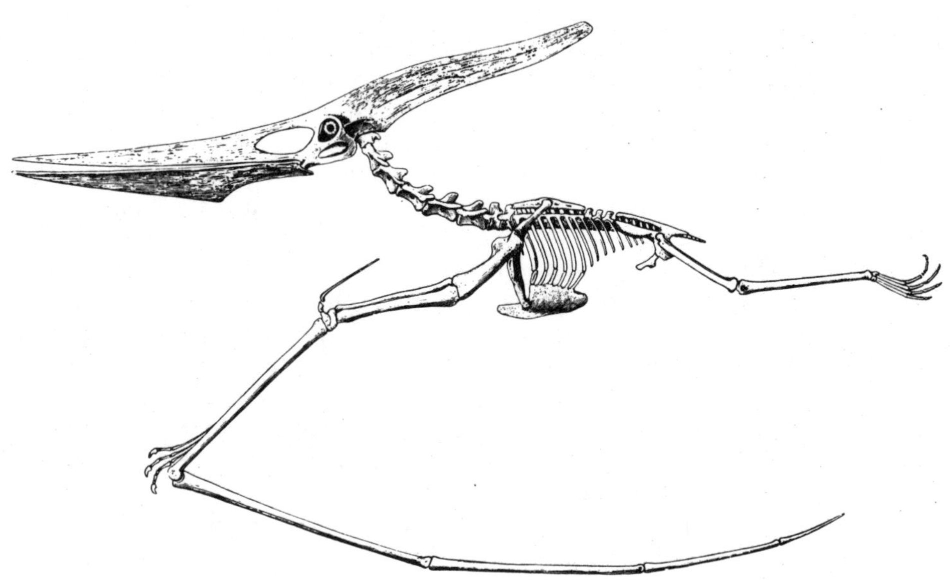

Skeleton of *Pteranodon*, a marine pterodactyl with a twelve-foot wing-span

which seemed quite at variance with the random changes upon which natural selection depends.

Marsh, having been elected President of the American Academy of Arts and Sciences, realised that in Wyoming was material for a life's work and launched a campaign to make Yale the greatest paleontological centre in the world. His rival was Cope. Often each man would find similar bones within a few days of one another and name them within a matter of hours. Cope took to telegraphing his papers to the American Philosophical Society to ensure priority. He accused Marsh of breaking up fossils he could not use himself to prevent others getting them and of hiring men to obstruct his rivals' work.

Cope was obliged to sell his house to cover his expenses. Marsh had to mortgage his house and take salaried work. When he died at sixty-eight he left only $200 in cash. Between them, however, they changed the face of paleontology. The 1871 edition of Lyell's *Elements of Geology* did not even mention the word 'dinosaur'. Marsh described thirty-four genera and made 'dinosaur' a household word.

This mighty battle is long forgotten and Cope is known to many students only for 'Cope's Law', which asserts, roughly speaking, that everything goes on getting bigger. As he had spent his life studying the reptiles which, starting about the size of a cat, ended up as the largest land creatures ever known it is not difficult to see how he came to this conclusion, especially as it was so much in keeping with the philosophy of nineteenth-century America. Alas, it is not generally true. Llamas and camels were once the size of hares, it is true, but the modern tiger is smaller than the sabre-toothed tiger of the last ice age. Marsupials, at first the size of a rat, became as large as rhinos – and then declined again. The horsetails of our ditches are tiny compared with the sixty-foot horsetails of the Carboniferous. And where are the giant snails of the early Cambrian or the giant oysters of the Tertiary?

However, there undoubtedly are quite a few cases where a family did steadily increase in size – the most obvious being the dinosaurs – and where this trend went to an extreme which must have been disadvantageous. The explanation usually offered by selectionists is that, wherever males fight for access to females, size pays: the bigger animal usually wins. The weakness of this argument is that steady size increases occur in classes where no fighting for the female occurs, for instance the ammonites, while in insects, some of which do fight, size remains small. The large animal needs to eat more than a small one, a handicap except where food is exceptionally plentiful. Since volume increases more rapidly than area, a large animal should be at an advantage in a cold climate and at a disadvantage in a hot one, since it has more heat to dispose of through a given area of skin. This is supposed to account for the large size of the mammoth, as against the modern elephant. But it does not account for the large size of *Diplodocus*, living in the warm, moist Jurassic; nor for the fact that small reptiles were equally successful. It does not account for the great size difference between hawks, falcons and eagles, all raptorial birds, living in much the same climatic environment.

If animals are sorted into size groups, it can be seen that there are far more small varieties than large ones, which suggests that, overall, size is not a plus. The fact is, natural selection really does not account for the varying size of different classes of animal. And it looks to me as if those animals which got progressively larger did so because the gears had stuck – because some genetic regulator had got switched on and could not be switched off. A large animal is not simply a scaled-up small one. It is 'ponderous'. Its larger mass calls for thicker legs, more powerful muscles, a stronger heart. It suffers more from inertia; it cannot get going, nor turn nor stop as easily as a smaller animal, and it is harder for it to hide.

Small wonder that many authorities have spoken of 'momentum' in evolution carrying a development beyond its adaptive norm.

Reconstruction of the mammoth, an animal somewhat larger than its modern relative the elephant

Reconstructed specimens of *Diplodocus*, a creature measuring up to seventy-five feet in length, together with a *Ceratosaurus*. Both lived during the Jurassic, about 150 million years ago

If gigantism is hard to explain in selectionist terms, the over-development of single organs is even harder. And this is very common, as we have seen: the tail of the lyre bird or the bill of the toucan, the tusk of the mammoth or the antlers of the Irish elk. In the sweeping way that evolutionists have, Simpson dismisses such cases as the bird of paradise and the peacock on the grounds that they are flourishing anyway. But to say that their displays are not disadvantageous does not make them advantageous, and unless they are positively advantageous they would not be selected.

In any case, there are harder cases to explain. The canine teeth of some prototheres such as *Smilodon* became so large that they could not close their mouths, while the tusks of others curved round as well as growing larger until they were of no value as offensive weapons; in fact they must have been in the way. Various desperate attempts to account for such monstrosities have been made. For instance, in 1910 Matthews tried to account for *Smilodon* by supposing it had a prey which vanished and that its huge teeth had been required in that connection and had remained; they were 'a lag in adjustment'. Even if this were true of *Smilodon* it would hardly account for the other cases, and clearly some more general explanation is needed. Perhaps the most plausible is that the size of the organ is genetically linked to size as a whole in such a way that doubling the animal's size more than doubles the size of the organ.

There seems to be some truth in this but it raises problems for genetics. Consider the horn of the titanothere, which is rather like that of a rhinoceros but bigger. The horn is wholly absent in the smaller specimens, but it gradually becomes gigantic in later and larger ones. If we plot the measurement of horn size and body length as a graph we get a straight line, showing that there is indeed a relationship as we might expect. But at small sizcs, the length of horn

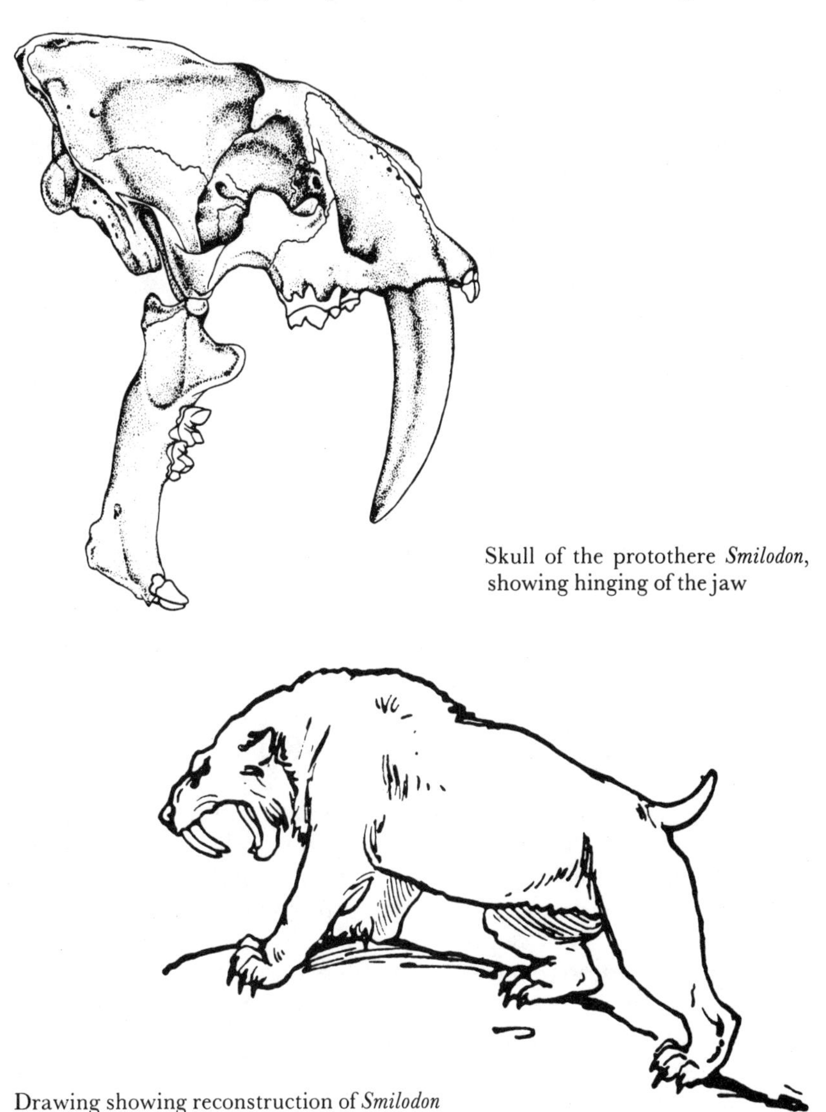

Skull of the protothere *Smilodon*, showing hinging of the jaw

Drawing showing reconstruction of *Smilodon*

Development of horn and body size in successive species of titanothere. In ascending order: *Eotitanops gregoryi, Palaeosyops leidyi, Brontotherium leidyi, Brontops robustus*

becomes negative! Genetics then has the problem of explaining how the size of something which does not exist is controlled by the genes which control size in general.

Here again, it is more plausible to believe that a trend is launched – perhaps because of its initial selective advantage – and then gets locked in, so that eventually an overshoot occurs.

Increase in size is, in any case, not the only kind of overshoot which occurs and the case for the 'jammed switch' theory is strengthened when we consider other kinds of overshoot. You will recall the oyster-like marine organism, *Gryphaea*, the shell of which coils more and more until it ends by preventing the unfortunate creature from opening its valves, so that it succumbs. This cannot be attributed to a 'lag in adjustment' or explained as a feature advantageous to the creature when young. It is just overshoot. So Verne Grant's sweeping dismissal – 'There is not a shred of evidence for the theory of orthogenesis in any of its forms', goes too far. Even if we agree to neglect those trends created by a change of behaviour (such as readaptation to marine existence) and those imposed by a change of environment (such as adaptation to a slowly cooling climate by growing a thicker coat of hair) there are still trends which call for explanation.

And when you come to think of it, trends are everywhere: the trend to the fish form, to the amphibian form, to the mammalian form; these are all prolonged shifts or trends. So, of course, are the regressions, as even Ernst Mayr admits. The evolution of whales from the Eocene zeuglodonts, like the evolution of birds from pseudosuchians, 'seems indeed as decisively directed towards perfect adaptation in the new medium . . . as if someone had directed the course of evolution'.

Professor Brough concludes that the attempts of the geneticists and people like Professor Simpson to explain orthogenesis away are 'not completely successful'. However, much better evidence of the presence of a directive trend in evolution is available, as we shall now see.

2 *Repetition Work*

It has repeatedly happened in the course of evolution that similar organs and even similar organisms have evolved on more than one occasion. For instance, legless forms, like snakes and worms, arose repeatedly from legged forms. The herbaceous plants arose again and again from woody ancestors. Each of the traditional sub-orders of rodents, from squirrel to beaver, arose several times from the generalised early rodents. An especially good example is afforded by the anteaters (which do not in fact eat ants but termites). Creatures which have specialised for such a diet by losing their teeth and developing a long, sticky controllable tongue exist today in South America, Australia and Africa; in Africa there are two versions. Each of the four kinds comes from a different evolutionary stock. Frank McCapra and Russell Hart of Sussex University, specialists in the ability of various fish, shrimps, bacteria, clams and dino-flagellates to emit light, say 'There is no doubt that bioluminescence arose many times in evolutionary history.'

The eye also evolved at least four times, but each time in a different form. Similarly the wing of a bat, which is a mammal, is very like the wing of a pterodactyl, a reptile which vanished millions of years before the bat appeared.

Chance has presented biologists with two remarkable opportunities to study such 'parallel evolution'. Vast earth movements, as we

Wing of a pterodactyl (top), a bat and (bottom) a bird

have recently come to realise, once split up the primeval land mass, which geologists call Pangaea. North America drifted off one way, Asia the other. Later an east-west split developed, separating South America from North America, on the one side, and Australia and Antarctica on the other. Only much later were North and South America rejoined by the Panamanian isthmus. In consequence, South America and Australia became vast natural laboratories in which the existing plant and animal stocks developed independently of those in Asia and of each other. Curiously enough, both evolved the marsupials, which carry their young in a pouch, e.g. the kangaroo. The kangaroo is unlike any Old World mammal but Australia harboured a marsupial which closely resembled the Old World wolf – known as *Thylacinus* – and another which was remarkably like our old friend the mole.

Much the same is true of South America. Indeed the porcupines of South America are so like Old World porcupines that some authorities have been driven to suppose that they got there by drifting across from Africa on a raft of vegetation! To take another case, South America harboured the liptoterns which are not unlike horses, the mesodon which looked like a rhinoceros, and the typotheres, some of which looked like rabbits while others resembled sheep.

While there were undoubted resemblances in some groups I am personally more struck by the fact that some of the creatures developing in these enclaves were quite strange. There was, for instance a rodent with enormous eyes, walking on its two back legs like a miniature kangaroo, and with a nose prolonged into a bony tube. So large were its eye sockets that there was no room for the attachment of the jaw muscles to the skull. It disappears suddenly from the fossil record, leaving the niche wide open. Again, there were the Groeberidae, called by G. G. Simpson 'one of the strangest' as well as 'one of the rarest' families. 'These highly specialised animals appear in the known record without any antecedents and then immediately – geologically speaking – disappear again for ever.' The whole fauna in which they appear, he adds, is peculiar. Sudden appearance and disappearance also marks the Necrolestidae, marsupials resembling moles. The 'evolutionary laboratory' of South America presents Darwinists with as many problems as it solves for them.

Hardest of all to explain is the fact that the earliest South American fossils known reveal only three orders of mammals, already much diversified: the marsupials, the hoofed herbivores and the Xenarthra or creatures with strange joints, e.g. the armadillos. 'Such a mix is not only peculiar; it is unique,' observes Simpson with some chagrin.

Evolutionists interpret these similarities as evidence for natural selection, commenting that similar environments have provided niches for similar life-forms which have evolved to fill them. But it seems to me the evidence could just as well be interpreted in the opposite sense. Why did Australia and South America, which are not particularly alike environmentally, both contain marsupials? And why did the Old World, which includes environments resembling those of both those continents, *not* produce any marsupials, except in Europe in the Eocene? Moreover, it is only a proportion of the forms which are similar. There are no kangaroos in South America, no armadillos in Australia, and no duck-billed platypi in the Old World. One might justifiably say that, in these isolated eco-systems evolution wandered off in new directions.

Struck by the similarities, some biologists have wondered whether the genetic system contains only a limited number of building plans.

Of course, the environment unquestionably imposes certain limitations. Creatures eating the same food require the same sort of teeth. Fish which migrate from salt to fresh water need more efficient kidneys; so do crustaceans which make the same move. Both, in fact, evolve longer kidney tubules. Such parallelisms are especially noticeable when lines revert to an earlier life-style. For instance when mammals enter the water they lose their legs and develop a powerful propulsive tail, just as the reptiles did in similar circumstances.

However, parallel evolution also occurs between different lines at the same epoch in time and some of these changes are much harder to explain.

The forerunners of the mammals were a group of animals known collectively as therapsids. Remains of them have been discovered as far apart as North America and Yunnan in China. Everett Olson of Chicago University studied them back in the 'fifties. He found that various lines seemed to be acquiring mammalian characteristics quite independently, in parallel. He listed fourteen changes, none of which appear in other reptile groups. These included modifications as diverse as the rotation of the semicircular canals of the ear and the initiation and development of a secondary palate separating eating from breathing, together with numerous changes in the bone structure and reduced thickness in the base of the skull. Some appeared to be adaptations for a specific purpose (e.g. the improvement of the ear) while others seemed to be more general (e.g. the thickness of the basicranium). These changes seemed quite unrelated to the very different niches which the several lines occupied.

How, he wondered, do you get whole suites of characters all appearing in different lines? Such facts, he wrote, 'have at times in my own mind and I am sure in that of others given rise to an uneasiness with regard to the current genetic radiative concepts of evolution'. And he wondered: 'Is there a gap in current theory which, whilst it may be correct, is not sufficient?' Examples of multiple-directional evolution are plentiful in the fossil record, not only among mammals but also among reptiles and fishes, but many paleontologists, especially the English ones, have been content to 'leave well enough alone' and get on with the task of description without asking awkward questions, Olson observed, rather acidly.

It was a question which had been asked already by Professor D. M. S. Watson.

In 1912, accident took Professor D. M. S. Watson to Newcastle, just after a visit to South Africa where he had seen a fossil of the newly discovered uranocentrodon. Visiting the Natural History Society, he discovered a remarkable collection of fossils: the Thomas Atthey collection.

Thomas Atthey kept a grocer's shop in a mining village near Newcastle, in the mid-nineteenth century. But what passionately interested him was fossils. He persuaded colliers to bring him what they found, paying them in packets of sugar and tea. Then, with enormous care and patience, using a pair of needles, he separated them from the rock in which they were embedded. A twentieth-century authority has called him 'the perfect preparator' and it is only recently that his methods have been improved upon. Between 1869 and 1878 he wrote a number of scientific papers which were ignored for thirty years. His grocery business failed and he was forced to sell his collection to the Newcastle Natural History Society in order to live.

This happy discovery launched Watson into a study of the labyrinthodonts, the main group of fossil amphibia, so called because of the strange folded structure of their teeth. They were the link between fishes and reptiles.

After examining many features of their structure, he wrote: 'There remain two features of the evolutionary story of amphibia which seem to be quite incapable of explanation on any adaptational basis.' These are the steady flattening of the head and forward part of the body and the progressive reduction of cartilage bone in the skeleton. 'It seems in every way likely,' he concluded, 'that this flattening . . . is brought about by internal factors not directly influenced in any way by the environment.' The gradual disappearance of cartilage bone, occurring both in marine and freshwater species, could not be due to shortage of calcium or potassium, he felt. 'It must depend on internal factors.'

How, he asked, 'can a group in which many basal stocks are of different habits pursue parallel evolutionary courses?'

Subsequently, Watson turned his attention to the origin of frogs, one of the notorious mysteries of evolution. The manner in which their limbs and muscles have been adapted for leaping separates them from every other vertebrate. The story of how he was able to solve the problem is worth telling, if only because it illustrates the slow and chancy nature of these studies. The fossil which was to prove the missing link between the frogs and what we now know to be their amphibian ancestors was actually discovered in the US as far back as 1865, by the untiring Edward Cope. But somehow it was lost and it was not until 1909 that another specimen was found. This was deposited in the Yale Museum, unrecognised and still largely covered with the kaolin in which it had been embedded. It was known simply as No. 794. There Watson found it, in the 'thirties. Carefully scraping off the kaolin, he realised he was looking at a specimen which could resolve the problem of frog origins. The curator, Professor Romer, on being informed, provided him with another and better specimen, which was christened *Miobatrachus romeri*, meaning 'Romer's less than a frog.'

Watson felt sure he now had the link back to the ancestor – a small, scaly creature known only from one specimen and entitled *Eugyrinus*. But he could not make the link forward, for there was no known specimen of the father of all the frogs. Not until 1937 did France's Professor Piveteau discover the remains of such a frog in the Lower Trias: he named it *Protobatrachus*. Watson then could prove his case.

Watson found among the frogs the same flattening of the skull, the same reduction of cartilage bone as he had found among the labyrinthodonts and noted that it continued uninfluenced by changes in the animal's habits.

Watson fully recognised that there were other changes occurring in the skull which *were* caused by natural selection, acting in conjunction with the two primary trends before-mentioned. The conclusive point was that, in different lines of frogdom, these secondary changes occurred at varying rates, so that eventually the lines became quite unlike, *despite the fact that all were subject to the same evolutionary pressures!*

The case against the all-embracing nature of natural selection could hardly be made more clearly.

Parallel evolution is even more widespread among plants than it is among animals. In the early 'thirties, F. E. Fritsch analysed the several classes into which the algae are divided. In five of them he found essentially the same basic structure and elaboration. In some cases only study of the cell contents under the microscope could reveal any differences in appearance. 'It is quite impossible to establish with any measure of certainty the systematic position of the simpler types of algae found in the fossil condition,' he wrote. 'The same kinds of events seem to have occurred quite independently in different phyletic lines,' observes Manchester University's C. Wardlaw, commenting on his work.

In the evolution of flowering plants particularly, he says, 'all the major trends seem to have evolved independently'. Let me give you a simple instance.

As you may have noticed, some flowers have petals which are separate from one another, like the rose and buttercup. In others, however, the petals are fused together, as in the primrose or the phlox. These two variations crop up independently, without obvious rhyme or reason. The same is true of many other features.

Even between the two great kingdoms of plants and animals there are certain parallels. Notably the fact that both rely for reproduction on an ovary fertilised by a male cell, be it pollen grain or spermatozoon.

Often the fossil record fails to reveal whether or not parallel evolution has occurred. Some botanists think the several classes of ferns evolved independently from algae; others think they diverged from a Devonian ancestor. 'Whether or not contemporary genetical interpretations alone can adequately account for the close similarities in evolutionary trends in unrelated groups remains an open question, at least for the uncommitted observer,' observes Wardlaw.

After the acceptance of Darwin's ideas, both plant and animal biologists set out on a grand attempt to construct 'genealogical trees' which would show how every group of plant and animal had descended from ancestors leading back eventually to one primordial cell, the ancestor of all. This was *the* task of paramount importance, affording a great centralising theme. Today, a century later, that ambition remains unfulfilled. Not only are the trees which they have sketched full of gaps and queries, but a growing awareness of how often evolution has occurred in parallel has brought about a reluctant acceptance of the idea that many evolutionary developments arose again and again, and this not only in respect of different organs, but of whole phyla. In the jargon of evolutionists, the idea of a polyphyletic origin has become acceptable even for the whole plant kingdom. Hence questions such as, did the mosses and ferns arise independently from the algae? may be unanswerable. All plants may not share a common ancestry.

This new mood of humility is summed up by C. Wardlaw in these words: 'Our knowledge of many aspects of plant evolution is still inadequate, not to say meagre . . . We still do not know – and indeed, there appears to be no firm opinion among botanists – whether the Plant Kingdom should rightly be envisaged as a monophyletic or a polyphyletic system. Nor do we know how to relate the more advanced groups to their presumed simpler progenitors, e.g. the bryophytes or pteridophytes to their presumed green algal ancestors; the pteridophytes to the bryophytes; the gymnosperms to the pteridophytes; or the flowering plants to the gymnosperms or the pteridophytes. As to the great contemporary group of the flowering plants, we know virtually nothing of its origin, or origins, and little more of the mutual relationships of its major subdivisions.'

Despite all this, most textbooks still offer students neat family trees of the pretended origin of plants and animals.

The orthodox evolutionists insist that 'Although many such parallelisms have not yet been adequately analysed there are no indications suggesting that unknown forces are at work.' That sounds to me like a wishful thinking. The same authority, the magisterial Bernhard Rensch, of the University of Münster, affirms 'We may state that the interpretation of parallelisms does not require the assumption of special autonomous factors of evolution.'

It may well be that *some* forms of parallelism are explained by the demands of a particular environment, but that does not justify the conclusion that all are so explained. The alternative possibility, which Rensch lightly throws off, is that similar genetic changes happened to occur in coexisting lines. That this should happen for fourteen characters in eight lines of descent, however, is so wildly unlikely as to be laughable.

Professor Brough is one of those who find the occurrence of such non-random changes indigestible. Since they are not random, he insists, they cannot be due to natural selection. Even paleontologists who take an extreme neo-Darwinian view admit the existence of these non-adaptive changes. It is absurd to call them 'pre-adaptation' as G. G. Simpson does; he might as well call them non-adaptive, Brough declares. 'There seem to have been evolutionary surges in the past . . . where large changes of organic form took place . . . [in which] natural selection played little part. There is plenty of evidence that during these evolutionary surges changes produced by mutations were not random but directional . . .'

To many investigators the existence of parallelisms has seemed strong evidence of a plan; to others it has suggested Lamarckist interpretation. For my part I do not find the fact that (for instance) the pterosaur and the birds evolved a rather similar wing surprising, or that the placodonts have an armoured body like the turtles from which they differ greatly. But I do think it calls for some reconsideration of the genetic assumptions on which selectionists rely.

3 Chicken and Egg

In an earlier chapter we came across cases where new structures seemed to have appeared *before* they were needed. If this really happens it completely explodes the theory of natural selection and we need no further evidence to undermine it. What we would need is a new theory.

One instance I mentioned was the change in the fins of fishes, from a form supported by rays to one stiffened by a system of bones resembling those later found in the limbs of animals. That is to say, an upper arm, a lower arm comprising two bones, a bundle of wrist bones, and a number of digits each made up of phalanges. This seems a rather arbitrary way of supporting a fin. The orthodox explanation is that fishes, having become trapped in rock-pools by the receding tide, wished to walk back to the ocean. To my mind, this suggestion is little short of ludicrous. As anyone who has ever looked into rock-pools knows, fish practically never get trapped in them, except perhaps for minnows which live in them quite happily until the tide rises again. Larger fish do not come so close inshore as to be trapped and most of them remain quite far out at sea. Even the minnows have more sense than to get trapped. We might concede that a few species of lung-fish, capable of surviving for months in the mud, found themselves in trouble and wished to look for a better hole.

Then there was the appearance of feathers on the archaeopteryx, seemingly fully developed. This time the official explanation is even more implausible: it is alleged that the feathers were not for flight but for warmth. Hair would have been quite sufficient for this purpose, as other animals have shown again and again, and much simpler to evolve. The interlocking barbules of feathers have no role to play in heat conservation though they are vital for flight.

Birds, incidentally, required a more efficient brain to handle the problems of flight and there is some evidence that their brain size increased ahead of time in reptiles like *Sphenosuchus* in the Triassic.

I have also mentioned the amniote egg pp. 63–4, which seems to have appeared on the scene while the reptiles were still mainly water-living, although its value was that it opened up new niches on land. However the fossil record is, as usual, incomplete.

Pre-adaptation can even be found almost at the start of the evolutionary story, for when some of the earliest single-celled creatures lost the power of photosynthesis (as I shall tell later) they needed a new set of enzymes to enable them to digest complex foods, and they needed them right away if they were to survive.

But I would like to tell you here of a much more clear-cut

instance of 'pre-adaptation' which is extremely difficult to explain away.

There are a number of species of frogs which have discovered how to climb trees, a manoeuvre for which they are not naturally well adapted, their powerful hind legs being evidently designed for leaping rather than clinging. One of the greatest experts on the biology of the amphibia who ever lived was G. Kingsley Noble, the explorer. He died in 1940. With his colleague M. E. Jaeckle he studied these frogs in the late 'twenties. On close examination, they noticed that the structure of their feet was quite unusual; the fingers were longer and there was an extra segment of cartilage. Under the microscope, differences in cell shape and arrangement, the distribution of skin glands and the arrangement of connective fibres were found. The cartilage pad gave the frogs a better grip on the tree – a sort of built-in climber's boot. A classical piece of adaptation, you might say, for ground-dwelling frogs do not possess these features.

But where the story becomes puzzling is that there are also a few lines of ground-dwelling frogs which never climb trees which do also have these adaptations. Noble and Jaeckle were quite clear in their minds that these were not frogs which had recently abandoned tree-climbing. They seemed more like frogs which were getting ready to be climbers. 'The conclusion seems obvious that the tree-climbing apparatus developed before the frogs began to climb trees,' they commented. In fact they went further and said: 'A detailed analysis of the many "marvelous adaptations" in the Amphibia will reveal, we believe, that *in most cases* the modification arose before the function.' (The italics are mine, for the statement would raise hackles on most biologists.) As one of them has said, 'I personally doubt the truth of the statement as expressed. But when I reflect that G. Kingsley Noble was one of the greatest experts on the biology of the Amphibia who ever lived, I do pause.'

Evolutionists like to confuse the issue, when discussing pre-adaptation, by applying the term not only to the kind of cases I have just described but also to cases where a feature evolved for one purpose turns out to be useful for another.

For a reasonable example of this, I give you Weddell's seal, which lives further south than any other mammal. It swims far under the ice and uses its massive teeth to cut holes in the ice through which to breathe, in places where there are no natural breaks in the ice. This enables it to escape the attentions of killer whales, which cannot pursue it so far without running out of breath. Killer whales are the commonest cause of death among other seals, whereas, not too surprisingly, breakage and wear of their teeth are the commonest causes of death in Weddell's seals.

I am quite willing to believe that, in this case, the seal evolved good teeth for the usual masticatory purposes and then evolved them further to cut ice. The trouble is that if you use the concept of pre-adaptation to cover cases like this, you end up by including everything: all adaptation is pre-adaptation since a feature can't be selected until it is there.

To say, as Douglas Futuyama does (he works at the State University of New York at Stony Brook), that the Australian rhea was 'pre-adapted' to slaughter sheep, when they were introduced into Australia, because it had a slashing great beak, is like saying that the human hand is pre-adapted to hold knives and forks. It is pathetic to trot out instances of this kind and then imagine that the real problems I have described have been satisfactorily dealt with.

4 A Summing Up

The evidence I have presented is only a sample from a mass of data, but it is enough, I suggest, to prove that natural selection is insufficient to explain all the features of the evolutionary story and to make it necessary to consider quite seriously the possibility that some directive force or process works in conjunction with it. I do not mean by that a force of a mystical kind, but rather some property of the genetic mechanism the existence of which is at present unsuspected.

In the preceding chapters we have seen at least a dozen areas where the theory of evolution by natural selection seems either inadequate, implausible or definitely wrong. Let me briefly summarise them.

(1) The suddenness with which major changes in pattern occurred and the virtual absence of any fossil remains from the period in which they were alleged to be evolving.
(2) The suddenness with which new forms 'radiated' into numerous variants.
(3) The suddenness of many extinctions and the lack of obvious reasons for such extinction.
(4) The repeated occurrence of changes calling for numerous co-ordinated innovations, both at the level of organs and of complete organisms.
(5) The variations in the speed at which evolution occurred.
(6) The fact that subsequently no new phyla have appeared, and no new classes and orders. This fact, which has been much ignored, is

perhaps the most powerful of all arguments against Darwin's generalisation.

(7) The occurrence of parallel and convergent evolution, in which similar structures evolve in quite different circumstances.

(8) The existence of long-term trends (orthogenesis).

(9) The appearance of organs before they are needed (pre-adaptation).

(10) The occurrence of 'overshoot' or evolutionary 'momentum'.

(11) The puzzle of how organs, once evolved, come to be lost (degeneration).

(12) The failure of some organisms to evolve at all.

And one might add here the failure of various conceivable patterns to emerge, despite the overwhelming tendency to diversify. (The curious absence of six-legged tigers.)

In view of these facts, the bland confidence of workers like Professor Rensch that all can be explained in terms of mutation and natural selection seems distinctly misplaced. And as we shall see when we come to inspect the behaviour of animals (and to a less degree of plants) there are inexplicable – or at any rate unexplained – similarities between widely different groups there too.

The pontifical assurance of the orthodox evolutionists has obscured the fact that the number of highly competent biologists who have voiced doubts, or actually declared themselves in support of directiveness, is quite large and has continued ever since Darwin launched his ideas. Darwin's contemporary Karl Ernst von Baer, one of the most eminent of nineteenth-century biologists, used the word 'purposefulness' in criticising Darwin's views. Looking through my notes I see that in 1940 (for example) the German biologist Professor von Huene concluded that there was 'a superior principle governing and directing the whole'.

Especially thorough was the work of the Swiss biologist H. Wehrli who, after studying the evolution of horses during the Miocene, declared; 'The process of transformation seems to be directed and not at all caused by random mutation and selection.' His studies of the marmot led him to the same conclusion. 'The phylogenetic transformation of the Alpine marmot cannot be interpreted as caused by random mutation and selection, either, and one definitely is inclined to assume immanent forces transforming the animal type in the direction which began initially.'

Then again K. Beurlen was impressed by the many cases of over-specialisation which had no adaptive value: 'It is absolutely clear,' he wrote, 'that adaptation and selection cannot be regarded as the essential principles of phylogeny, since (in these cases) it proceeds from adapted to non-adapted types and it produced disadvantageous organs from useful structures. It is an autonomous regularity of its own . . .'

And in our own day, Sir Alister Hardy, formerly the Linacre Professor of Zoology at Oxford and now Professor Emeritus, after considering the problem of homologous structures and the problem of animals with the same form in different environments, reflected: 'I may be unduly sceptical, but I cannot help wondering if there is not something else concerned with the evolutionary process that we do not understand . . .'

When we come to consider the genetic aspect I shall try to indicate what I think that something is. But before doing so, there is one more topic which merits our attention: the origin of species.

CHAPTER SEVEN

The Unsolved Origin of Species

1 The Paradoxes

Since Darwin's seminal work was called *The Origin of Species* one might reasonably suppose that his theory had explained this central aspect of evolution or at least made a shot at it, even if it had not resolved the larger issues we have discussed up to now. Curiously enough, this is not the case. As Professor Ernst Mayr of Harvard, the doyen of species, studies, once remarked, the 'book called *The Origin of Species* is not really on that subject', while his colleague Professor Simpson admits: 'Darwin failed to solve the problem indicated by the title of his work.'

You may be surprised to hear that the origin of species remains just as much a mystery today, despite the efforts of thousands of biologists. The topic has been the main focus of attention and is beset by endless controversies. In the early years of the century, species formation was not regarded as presenting any problem. Everyone was preoccupied with debating how the variations upon which natural selection might act, arose. It was the broader evolutionary changes which monopolised attention – could the accumulation of minute variations account for the major steps in evolution?

But in the last thirty years or so speciation has emerged as the major unsolved problem. The British geneticist William Bateson was the first to focus attention on the question. In 1922 he wrote: 'In dim outline evolution is evident enough. But that particular and essential bit of the theory of evolution which is concerned with the origin and nature of species remains utterly mysterious.' Sixty years later we are if anything worse off, research having only revealed complexity within complexity. Only an occasional optimist like Verne Grant, of the Rancho Santa Ana Botanic Garden, feels able to say: 'Some of my readers may feel, as I do, that we are now close to having a satisfactory general theory of speciation.' More biologists would agree with Professor Hampton Carson of Washington University, St Louis, when he says that speciation is 'a major unsolved problem of evolutionary biology'.

Let me start by pointing to three problems of a general kind raised by this concept before plunging into the subject in more detail.

First, if a line of organisms can steadily modify its structure in various directions, why are there any lines stable enough and distinct enough to be called species at all? Why is the world not full of intermediate forms of every conceivable kind? (In point of fact, one or two plants – such as the willow – do behave in much this way. In contrast, the gingko tree exists in only one species.)

Until the eighteenth century it was generally held that there were certain forms which God had created once and for all and which remained fixed: pigs and horses, potatoes and roses; Plato had established this idea. Even more than Darwin, Lamarck's ideas dissolved this comforting fixity. Nevertheless, the system of classification by genus and family, which goes back beyond the great eighteenth-century classifier Carl Linnaeus to the sixteenth-century Swiss botanist Gaspard Bauhin, who devised it in 1596, continues to work very well today. So perhaps there is something in the idea of species after all.

The second problem is this. As we saw in the case of the darkening moths, and as we shall see again in more detail, the evolution of a new species can occur comparatively rapidly, perhaps in a space of two or three thousand years. For instance in a small lake which became isolated from Lake Victoria in Africa only 4,000 years ago there are now five endemic species of cichlid fish. Likewise, in the Philippines, Lake Lanao contains no fewer than fourteen species of endemic cyprinid fishes, showing curious modifications of the teeth and jaw – quite unlike the other members of the vast family – which must have developed since the lake was formed some 10,000 years ago.

Despite this, many species and even whole families remain inexplicably constant. The shark of today, for instance, is hardly distinguishable from the shark of 150 million years ago. And this constancy is seen at higher levels too: birds vary widely in size, shape, colouring, song and habits but are still substantially similar to the birds of the early Tertiary.

According to Professor W. H. Thorpe, Director of the Subdepartment of Animal Behaviour at Cambridge and a world authority, this is *the* problem in evolution. He said in 1968: 'What is it that holds so many groups of animals to an astonishingly constant form

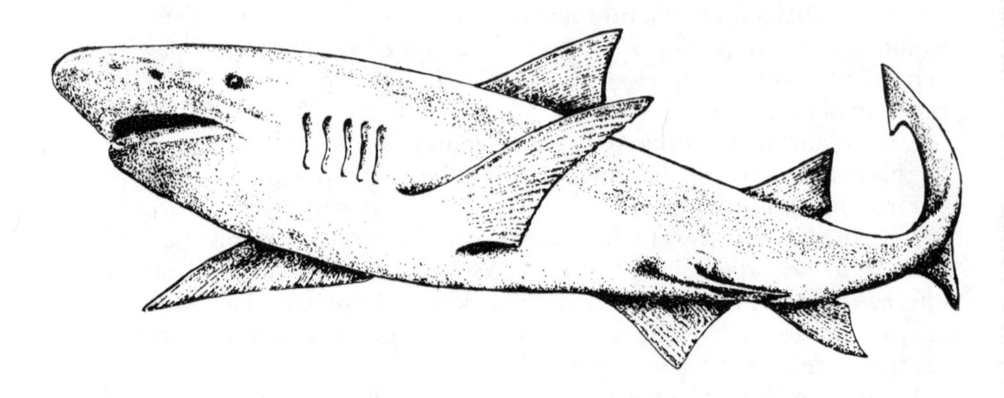

The shark has shown little evolutionary development over several millions of years

over millions of years? This seems to me *the* problem [in evolution] now – the problem of constancy, rather than that of change.'

But the theory of natural selection embodies a contradiction even more awkward and fundamental than these. The theory declares that organisms will become modified so as to function better in the circumstances in which they live. Earlier Darwin had put it like this: '. . . the varying offspring of each species will try (only a few will succeed) to seize on as many and as diverse places in the economy of nature as possible.' So far, so good. Or, as biologists put it, they will adapt to the niche in which they live. In cold conditions, furred animals will evolve a thicker coat. In a dry climate the skin will become more leathery to avoid water loss. A bird living where there are mostly hard seeds or nuts to eat will evolve a nutcracker type of bill. Why then do we find wide variations of type where the environment is identical?

I have already mentioned, as an example, the *Podostemaceae*, which Willis studied in Ceylon. He found forty genera, comprising 160 species, living in rapidly flowing water. Though the conditions were very uniform, he found an amazing variety of form 'greater than any other family of flowering plants'.

The word 'adaptation' is therefore ambiguous, since it is cheerfully used to mean fitting better into a niche and, on the other hand, modifying to fill some different niche as when we say, for example, that the mole is adapted for a life underground. But if 'adaptation' is a trick word, the word 'niche' is more troublesome still. The idea is attractive. One can see that a bare rock in the Atlantic does not ordinarily provide a niche for mammals, but can form part of the niche occupied by birds which live on fish. The trouble is that we only know a niche exists when we see it occupied. Some bacteria have learned to digest plastic materials, thus revealing the existence of a niche. Some areas of the deep oceans are so highly radioactive, others so loaded with heavy metals, that one might have thought they offered no niche, yet certain species have evolved mechanisms for coping with these stresses and live in them. We cannot say when a niche is full or how far it extends. Niches may vanish or appear, sometimes rapidly, as when a volcanic eruption occurs, sometimes very slowly. The only value of the word 'niche' is linguistic. It makes it possible to write about adaptation without defining the circumstances to which adaptation is occurring. And as the word adaptation avoids the need to define the process, it becomes possible to write a great deal without saying very much.

But if evolution adapts creatures to the niche they are in, how does it come about that some of them move to a different niche and begin to differ from the parent stock? Natural selection should, one would have thought, produce similarity rather than diversity. As Professor Richard Levins complains, natural selection 'does not explain how occasionally a breakthrough into a new niche or new mode of adaption [*sic*] takes place'.

Professor Theodosius Dobzhansky, a leading expert on evolution and highly orthodox, nevertheless becomes quite plaintive on the subject. 'To use Drosophila as an illustration: the more than 600 known species of this genus all have three orbital bristles on either side of their heads, and the anterior of these bristles is always proclinate (bent forward) while the others are reclinate (bent backward). Now why should this character be retained so tenaciously in so many species? Is it really important for the flies of this genus to have one proclinate and two reclinate orbital bristles?'

Adaptation, in short, is distinct from speciation. So in this chapter I shall look separately at speciation and adaptation, after further consideration of the notion of species itself.

2 *Elusive Species*

The people who decide what, in any given case, constitutes a species are called taxonomists. Human beings, it is said, tend to be 'lumpers' or 'splitters'. Taxonomists are splitters, delighted to find some trifling difference between almost identical objects, whereas lumpers prefer to point out what differing things have in common.

Many of the distinctions made by taxonomists are so trifling as to be meaningless: tiny differences in the banding of the shell of a snail, for instance, which have no biological significance. The French taxonomist Locard classified the freshwater mussels of France into no fewer than 251 species on the basis of their shell forms and colours. Today all 251 are regarded as a single species. Another man recognised 200 species of snail in Hawaii in 1905, whereas a three-man team who went there seven years later put the number at forty-three. Darwin's finches themselves have been the subject of this treatment, having been classified into over thirty species by a taxonomist who commented later that he might as well have called them all one species.

We should therefore perhaps take several grains of salt with the following figures, compiled by Verne Grant, which convey nevertheless an impression of the prolific way in which life has speciated. So far, some 8,600 species of birds have been described and about 3,700 mammals. But the number of fish listed is much higher, at 20,000 out of an estimated 40,000 believed to exist. Known insects number an amazing 850,000, but probably fewer than one-fifth of all those which actually exist have been described – maybe as few as one-tenth. Even worse are plants: the number of known flowering plants has been put at 286,000 and about 4,000 more are classified and described by taxonomists every year. Known fungi total over 40,000.

Adding it all up, Grant arrives at a grand total, for organisms now extant, of 1.6 billion. But countless species which once existed are now extinct. Estimates of the total number of species which ever existed range from 1.6 billion to 16 billion.

He comments 'The diversity of life is inconceivable.' But one could also say the opposite, for all these numerous species fall into some thirty main groups, the phyla, which seem to remain rather constant for indefinite periods.

One might well ask: why are there quite so many species? One might well suppose that natural selection would eliminate all but a few highly efficient forms, which would occupy the most favourable niches and dominate the scene.

However, the concept of species is bedevilled by a more immediate phenomenon which provides a headache not only for taxonomists but for the biologists who try to account for the evolutionary process. Namely, that the phenotype or physical appearance of many species can vary widely in different circumstances, even though the genotype, or genetic constitution, remains constant. The first botanist to study this phenomenon scientifically was an Austrian, Anton Kerner. As long ago as 1894, he collected seeds from lowland plants and sowed them at a height of 7,200 feet in the Alps. The resulting plants, he found, had shorter stems, smaller leaves, fewer and smaller flowers; the flowers were borne nearer the ground and the plants were brighter in colour both in leaf and flower than the plants from which they came.

Conversely, 'As soon as seeds collected in the Alpine region were again sown in the beds of the Innsbruck or Vienna Botanic Gardens, the plants raised from them immediately resumed the form and colour usual to that position . . . In no instance was any permanent or hereditary modification in form or colour observed.'

Just before the war a trio of American experimenters carried out a variation on this experiment. They dug up a plant of *Potentilla glandulosa* in the Californian coast ranges and brought it up in four different environments: dry and sunny, damp and sunny, dry and shady, moist and shady. It produced four strikingly different phenotypes. The plants grown in shade were taller and had broader leaves; the plants grown in moist conditions were more vigorous. Similarly, studies of black mustard show that it varies in at least ten respects according to whether it is grown in open or cultivated fields.

Many animals change seasonally. Not only do Arctic hares and foxes change the colour of their coats from brown to white as summer changes to winter, but even the tiny *Rotaria* in the village pond change shape.

The most striking example of phenotypic variation, I suppose, is the difference between the two sexes. There was a classic instance of how far this can go in the nineteenth century, when the king parrot which is brilliant red was classed as one species and the female, which is yellow, was classed as another. The taxonomist concerned was deeply embarrassed when the real relationship emerged. The ornithologist Ernst Mayr, who made an extensive study of birds in the Pacific, reported that the small whistler, the male of which species is normally coloured black, yellow, white and olive, while the females are a dull yellowish ochre, has lost its colour in the Rennell Islands and Norfolk Island and is indistinguishable from the female. He even found a race in the Solomon Islands and Samoa where the female had the brighter colours and general appearance of a male while the true males would have been taken by a casual observer as females.

In contrast with the cases where genetically identical organisms look different are cases where genetically distinct organisms look the same. For instance, the annual flower clarkia which decorates many cottage gardens. *C. lingulata* is indistinguishable from *C. biloba* by petal shape, but on examination is found to have a substantially different chromosome structure. The number of chromosomes is actually different, while parts of one chromosome have been 'translocated' to a new position, or inverted.

These phenotypic variations can fool taxonomists and one can imagine that they could fool paleontologists even more readily, but the reason I mention them is of more fundamental importance. How can natural selection act on a plant which has already assumed the form appropriate to its situation by its own efforts? No one knows. The fact that we have no idea how organisms make these adjustments does not help.

At this point the reader has to take a deep breath and swallow the fact that biologists, totally ignoring the taxonomists, use a quite different and far more rational method of defining a species. In biology, a species is a group of creatures which are interfertile – that is, they mate with one another and produce offspring. Thus two groups which are not interfertile constitute different species. The significance of this definition, of course, is that a new heritable feature, if advantageous, developed by one member will tend to spread through the group, but will not spread to other groups. Accordingly such groups will tend to diverge.

Sometimes we find cases where group A mates with B, and B with C but C does not mate with A. If B drops out, we shall be left with two species instead of one. Thus the lesser black-backed gull and the herring gull occupy a range like a ring around the North Pole. The British lesser black-backed gull, a handsome bird with a dark mantle of feathers and yellow legs, grades into the Scandinavian version, which in turn gives way to the Siberian Vega gull, which has dull, flesh-coloured legs and grey feathers. This in turn grades into the American herring gull, which grades into the British herring gull, with its pink legs and pale grey mantle. The species at the beginning of the chain differs from the one at the end both in appearance and habits. The former breeds on moors inland, the latter on cliffs, to name only one distinction. They behave as different species. Similar but more complex patterns of relationship have been charted for the Californian salamander, and for certain snakes, among others.

Just now I mentioned Ernst Mayr, calling him an ornithologist, which is how he started his career; but this does him much less than justice, for he has become the dean of evolutionary theory. Born in Germany, where he rose to be Assistant Curator in the Zoological Museum in Berlin, he left Germany and became an American citizen in 1931. He has received almost every medal it is possible for an evolutionary theorist to acquire, including the Wallace-Darwin medal itself. His views must therefore be listened to with attention.

In 1942 he propounded a theory of how species formation occurs which has become a dogma and appears in countless elementary textbooks and popular accounts of evolution – but it is a theory which has recently been challenged and is, I think, almost certainly incorrect. Mayr claims that all speciation occurs when, and only when, two populations of creatures are geographically separated, with the consequence that they cannot interbreed. Hence if one group undergoes an advantageous mutation it spreads through that group but cannot spread to the other. And conversely. When populations live on islands some distance apart it is obvious that they cannot interbreed and they are likely to drift apart genetically. Island populations have therefore been the object of careful study by biologists trying to catch speciation occurring.

For instance, Hampton Carson, a geneticist at Washington University, St Louis, made a close study of the fruit-fly, *Drosophila*, in Hawaii in the 'sixties and early 'seventies. The Hawaiian group offers special advantages for evolutionary studies, since the island of Hawaii itself was formed only 700,000 years ago, while to the north is Maui, formed at least 1.5 million years ago. Hence investigators know how much time was available for evolutionary changes to take place. Carson found three species of *Drosophila* of a sort known as 'picture-winged flies' on Hawaii which genetic studies showed must have come originally from Maui, where there are many species of this fly. He concluded that there had been only one 'founder event' – that is, a single gravid female had arrived in Hawaii by some unknown means and had given rise to a species which had diverged into three. Further studies suggested that, in the four main islands of the Hawaiian group there had been at least twenty-two such colonisations.

With the penchant for jargon which afflicts biologists, this rather simple idea was dubbed 'allopatric speciation'. If new species arise in groups occupying the same territory, it is termed 'sympatric speciation'. And if the areas overlap but do not coincide, they speak of 'parapatric speciation'.

So obvious did the need for geographical separation as a springboard for speciation seem to most biologists that one of them, W. H. Blair, could say in 1964 that geographical separation was 'so obvious a requirement as to need no further discussion'. Certainly, islands offer favourable conditions for speciation, such as the presence of empty niches (i.e. absence of rivals). Absence of predators may help especially while the group is small, and modifications will spread more rapidly through a small group. However, it is also true that small groups may easily breed out valuable genes. They easily perish.

But to claim that speciation mostly or always depends on isolation is quite ludicrous.

It is obvious that species often arise in circumstances where such isolation is highly unlikely or impossible. Lakes provide a particularly convincing example. Divergent species exist in the majority of lakes: for instance, Lake Baikal contains 239 species of amphipods alone. In the 'sixties doubts about Mayr's theory began to be expressed. Entomologists were particularly dubious, for there are usually countless species of insects in any given area, yet they speciate. However, Mayr defended his theory aggressively and developed it with so much erudition that many zoologists were convinced, especially those working with vertebrates.

When opponents pointed to seeming instances of sympatric speciation, Mayr grandly declared: 'The hypothesis is neither necessary nor supported by irrefutable facts.' He complained that the same old arguments were cited again and again, no matter how decisively he had disposed of them. His supporters argued that populations which now overlap were originally isolated but later extended their ranges until overlapping occurred. It seems unlikely that this occurred in every case – and Mayr certainly could not prove it. The most parsimonious assumption is that some species, at least, have always occupied identical or overlapping territories. Today many instances are known.

Biologists therefore began to look for other possible circumstances which would prevent two groups of organisms from interbreeding. There are plenty. Insect biologists pointed out that when the smell-receptors of females are removed they become indifferent to males; that the pheromones by which the male attracts the female might be species-specific – i.e. that a male would only attract females of the same species. This turned out to be true in some cases, not in others. A much more general way in which two groups can be functionally isolated is if their periods of sexual readiness do not coincide. Plants, for instance, cannot pollinate each other if their flowering times are different. Sometimes refusal to interbreed occurs for no discernible reason. There is, for example, in California a plant known as *Gilia* or the prickly phlox. Verne Grant has studied it in detail. Among many other observations, he found that of four varieties, three bred together freely while a fourth did not.

This reminds us that many animals and many plants form hybrids, the offspring of which are often, but not invariably, sterile. Why some species are more open to hybridisation than others is not understood. When hybridisation occurs in plants, the results can be very confusing to the taxonomist. It is also confusing to the biologist, who would like to know why some species find an advantage in hybridisation when the majority find it pays to be exclusive.

Population geneticists have been asking the wrong questions, as

Australia's Michael White puts it. Too many studies over the past decades have been designed more to exemplify Mayr's theory than to test it. And he cites approvingly the conclusion of a group of geneticists at the University of Texas: 'Speciation phenomena depend primarily on behavioural traits, such as those which determine reproductive isolation or a particular manner of niche exploitation and co-existence.'

3 *Speciation on Trial*

If you would really like to see the speciation process running mad, I suggest you visit the great lakes of Africa – Lake Malawi, Lake Tanganyika and Lake Victoria in the African Rift Valley. Each of the first two is some 350 miles long, a broadening of the upper Nile; the last is almost rectangular and nearly 200 miles from north to south, the world's third largest lake. Each of these lakes contains more species of fish than any other lake in the world and almost all of them belong to the family Cichlidae. There are believed to be 126 species of cichlid in Lake Tanganyika, and more than 200 in Lake Malawi, no species being common to both lakes. They are bony, perch-like fishes, up to three feet in length, with only one nostril and in some cases a bulging forehead.

One of the cichlid family of fishes, *Cichlasana eitrinellum*

All these species vary widely in the structure of their jaws and teeth, and are noticeably different from cichlids elsewhere in this respect. One species, for instance, has closely set teeth that form a scraper. Others have rows of fine, movable teeth which comb large algae from the rocks. They avoid competition by specialised feeding, living on algae, invertebrates, plankton, fishes and molluscs, and even leaves. The teeth are specialised for each of these foods. Some have evolved extraordinary feeding habits, such as *Genyochromis* and *Corematodus*, which feed on the scales of other cichlids, or the horrible *Haplochromis compressiceps* which darts at larger fish, snaps and withdraws, leaving behind an empty eye-socket. There are also fin-biters. Much of their evolutionary success is due to the large number of food sources they have found ways to exploit.

Among their unusual characteristics is the habit of carrying their eggs, and subsequently the young, in their mouth – not just one or two but sometimes forty or more. Both sexes show this behaviour and the father often collects straying young in his mouth and conveys them back to the mother, whereupon he spits them out again.

G. Fryer (of the Freshwater Biological Association) and T. D. Iles (of the Fisheries Laboratory at Lowestoft) call them in their magisterial book *The Cichlid Fishes of the Great Lakes of Africa*, from which I have derived these facts, 'a particularly remarkable group of fishes', and go on to say, rather daringly, 'We believe that cichlids are intelligent fishes: not in the limited sense that they can learn a particular skill but that they betray some of those attributes which make man the successful animal that he is. They exhibit an awareness of their surroundings, a definite alertness and what we might inadequately describe as inquisitiveness.' They go on to cite from Konrad Lorenz the story of a male cichlid which was gathering young in his mouth, in order to take them to their mother, when he saw and engulfed a worm. Immediately after, he noticed one of his offspring, swam after it, and also took it into his mouth. He now had in his mouth two objects, one to be swallowed, the other to be guarded carefully. After a pause of several seconds, presumably for thought, he spat out both, swallowed the worm without haste, and then picked up the young fish and conveyed it to the mother.

Cichlids often sham being dead, lying on their side until prey approach. Others – natives call them the 'crafty ones' – burrow in the sand to avoid the nets. They have an elaborate language, consisting of very specific changes of body colour. For example, the appearance of five black stars on the side of a female means 'I am guarding my young and will not tolerate interference.' A black mark and three white patches means 'I am busy spawning.' Nine grey patches, in contrast, mean 'I am frightened and am taking shelter.' Two black bars mean 'I am frightened and have nowhere to hide.' Fin position is also significant, and they appear to emit sounds for communication purposes, as well as chemical substances, such as a 'fright' warning.

Colour changes in some species can be astonishingly rapid. Thus in *Pseudotrophaeus auratus* the female displays three very distinct black horizontal bands on a gold background, while the male has three electric blue bands on black. When a male was caught and dropped in a bucket it instantly assumed the female colouration, except that the lowest band remained black, and a small part of one band became gold. As it recovered from its alarm, it slowly resumed its normal hues. 'The significance of this remarkable behaviour remains obscure,' Fryer and Iles remark.

Some of them build 'cities' of 200 or more nests, almost touching. Like man, but unlike most animals, they breed all the year round; and they are monogamous.

When you have had enough of looking at the specialist cichlids, you can turn your attention to *Tilapia*, another cichlid which makes up about half the fish population of the lake, and which is as unspecialised as the other cichlids are specialised. *Tilapia* shows amazing plasticity. It can alter its size, growth rate and fecundity to suit the current conditions of food-availability, etc. If food is short, it just produces smaller offspring, and these will grow on to normal size if more food becomes available.

Tilapia is tough and even survives successfully in Lake Magadi, a hot soda lake with an extremely high salt content, as well as in brackish and coastal waters. Unlike the other cichlids, it hybridises readily. The Egyptians recognised its remarkable qualities and, though they ate it, regarded it as sacred.

Looking at the conventional account of speciation in cold blood it seems perfectly obvious that there is something missing. Take, for example, the inexplicable variations in the speed with which it occurs. We have several clear-cut instances of speciation taking place very rapidly. I have already mentioned the five species of cichlid which appeared in Lake Nabugao in 4,000 years (see p. 141). Insects display even more rapid speciation. Thus in Hawaii there are at least five species of the moth *Hedylepta* which feed only on bananas, in contrast to their nearest relatives, which feed only on palms. Since bananas were only introduced into Hawaii by the Polynesians about a thousand years ago, the banana-eaters must have evolved from their palm-eating ancestor since then. Since they have also changed noticeably in appearance there is no doubt about there being a species difference.

In contrast, organisms like the blue mussel, the mountain lion or the brown creeper show little sign of speciation.

Then again, why do some families produce an enormous range of species, other not? There are countless varieties of juniper, but only one gingko. Let me digress for a moment on the incredible variety of orchids: it bears thinking about.

'Beings in a state of nature vary very little,' said Darwin – a strange remark, seeing that he was familiar with the orchids. This is the largest of all plant families, comprising between 20,000 and 35,000 species – no one has managed to count them all. In addition to the familiar epiphytes, living on host trees, they appear as shrubs, vines and even grasses. They are found from the edge of the Arctic all the way to Antarctica. They live in bogs and deserts, Alpine valleys and tropical rain forests, and tropical plains and high mountains. Obviously they have adapted with supreme success while still remaining clearly within the bracket of orchid.

In doing so they have produced a wild profusion of forms. One genus is popularly known as the swan orchid, another as the monkey face, another as the little bull. The seventeenth-century German botanist Jacob Breynius, one of the first to describe orchids in any detail, wrote 'If nature ever showed her playfulness in the formation of plants, this is visible in the most striking way among the orchids . . . They take on the form of little birds, of lizards, of insects. They look like a man, like a woman, sometimes like an austere sinister fighter, sometimes like a clown who excites our laughter. They represent the image of a lazy tortoise, a melancholy toad, an agile-ever-chattering monkey.' The shape of their flowers is so varied, he concluded, that one can imagine almost anything.

The late Oakes Ames of Harvard University, a leading orchid expert, felt that the orchids represented some kind of culmination in the evolutionary process, a view with which some others agree. They also have botanical peculiarities, including combined male and female sex organs; some can fertilise themselves by digesting away part of their own structure. Some of them produce more than a million seeds per plant.

When I consider the multifariousness of orchids I cannot escape the feeling that the process of natural selection – although it may well have been at work – is insufficient as an explanation and sometimes I wonder if the orchids did not achieve their success in defiance of its rules.

In Madagascar there is an orchid rejoicing in the name of *Angraecum sesquipedale*, the nectary of which takes the form of a spur eleven and a half inches long. When it came to Darwin's attention he predicted that a moth would be found with a proboscis capable of

In common with many species of orchid, the lady's slipper derives its name from its unusual shape

being extended to eleven and a half inches, and sure enough such a moth was found and christened *Xanthopan morgani praedicta*, because Darwin had predicted it.

Gavin de Beer cites this as a classic example of adaptation, saying that the longer the nectary, the more pollen would rub off on the moth, thus improving the species' chance of surviving, while the moth's long tongue would enable it to reach the nectar. This casual *ex post facto* argument is typical of the way in which evolutionists evade the difficulties of the evidence. I can see many disadvantages in such an arrangement. The orchid deprives itself of the services of all other moths, and if *Xanthopan praedicta* was hit by an epidemic *Angraecum sesquipedale* would probably become extinct. Moreover, the coordination of the two evolutionary processes has to be precise over some prodigious period. If the orchid lengthens its nectary, the moth – being unable to get any nectar – will cease visiting it and the orchid will remain unfertilised. I don't know what happens if the moth pulls ahead of the orchid in this strange race; perhaps it shortens its proboscis again. The whole thing seems so incredibly chancy one wonders if it wouldn't have been better never to have started. After all, other orchids and other moths get along very well without this co-evolutionary piece of trapeze work. It looks like showing off.

Biologists, as I said, use the term niche to describe the circumstances in which an organism manages to survive.

One of the odd features about niches is how many remain empty. For instance there are plenty of sea snakes in the Pacific and Indian oceans; why are there none in the Atlantic? Or again, why are there no blood-drinking bats in Africa? There are plenty of blood-drinking bats in the tropics of the New World, and fish-eating bats too. A quite recent case is the cattle egret, which lives on the insects stirred up by grazing cattle throughout the Old World. In the 'thirties these birds reached North and South America, where they are thriving. Even before there were cattle in North America there were bison, which would have provided the required conditions. It seems that ecological vacancies may persist for millions of years without any species evolving to occupy them.

Another unsatisfying feature is the failure of unwanted characters to disappear. Flightless birds, for instance, retain the orienting abilities they no longer need or use.

One factor for which little allowance is made is the behaviour of the organism. It has been shown that dark limpets tend to move towards dark backgrounds and light ones to light, in contrast to Kettlewell's moths, which adjusted themselves to the background rather than the background to themselves. The creature which thus avoids the pressure of selection will, presumably, remain unchanged.

Talking of moths, while *Biston betularia* managed to darken and lighten itself to suit the available backgrounds, T. G. Sargent, of the University of Massachusetts, found quite different behaviour in *Phigalia titea*, which preferred light backgrounds all the time and usually settled on white oaks. On them, the darkened forms (about 20 per cent of the population) were very noticeable. Yet they survive well. Puzzled, he notes: 'In such a circumstance, strong selection pressure for a more appropriate background might be expected.'

Evolutionists seem agreed that speciation occurs *before* adaptation. Thanks to some genetic shake-up, organisms evolve a new structure which then enables them to occupy a new niche or function more effectively in their existing niche. They then may adapt to the new situation. 'Speciation is not a route to improved adaptation,' concludes Douglas Futuyama. 'The time honored diagrams of evolutionary change are probably wrong.'

With this in mind, let us look at the extraordinary flexibility of adaptation, much harder to explain than speciation.

4 *Why Fishes Are Silvery*

Living things have adapted themselves to exploit an extraordinary range of conditions and have evolved mechanisms of unbelievable intricacy in order to do so. Life can be found at temperatures as high as 150°F in hot springs and also below the freezing point of water. It can learn to tolerate heavy metals which are normally poisonous to living things, and even high levels of radiation. It is found in oxygen-free atmospheres, in total darkness, under enormous pressures, in arid conditions, even in the gut of other creatures or inside living cells. But the strange thing is that while some organisms adapt to such unfavourable conditions, others wholly fail to do so.

However, adaptation does not serve merely to make extreme conditions tolerable. It also assists creatures to survive more easily in normal conditions. 'Can a more striking instance of adaptation be given,' Darwin exclaimed, 'than that of a woodpecker?' This bird has two toes turned back to front, so that it can grip the bark more firmly, a stiffened tail so that it can prop its body against the tree when pecking, a strong beak and a very long tongue for extracting the insects at the bottom of the hole.

In actual fact there are many even more striking instances of adaptation than the woodpecker. Consider, for example, the silvery appearance of fishes. If you have ever attempted skin diving you will have admired the shimmering reflections of a shoal of fish. Fish are mostly dark above and light below which helps to conceal them from any predator above or below them – in itself a neat adaptation. Seen from the side, their silvery appearance effectively conceals them in water which is full of light refracted by ripples on the surface. Divers often report that a school of fish has approached quite close before they spotted it.

How is this elusive appearance achieved? It is due to their ability to secrete many millions of tiny nitrogenous crystals in layers on their skin and scales. There are a million of these minute mirrors to the square centimetre. By themselves, they reflect about 25 per cent of the incident light, but a much higher figure – about 75 per cent – can be attained if they are arranged in layers, alternately with cytoplasm, provided that the layers are separated by exactly one-quarter of the wavelength of the incident light. For green light, this means about seven-millionths of a centimetre. And this is precisely what evolution has arranged. How can natural selection achieve such a uniquely precise solution?

Many sea creatures are transparent and would not gain from being silvered. But some vital organs are inevitably opaque, as is blood, and in such cases the organs alone are silvered. Again, some deep-sea fishes carry light-organs backed by silvered reflectors, which project a beam of reddish light. The eyes of such fish are equipped with orange-red filters! Since fish are normally only sensitive to green light, this enables them to see without being seen, an anticipation of the modern soldier's sniperscope.

It is, I suppose, just credible that a series of chances could have evoked such changes – so now let me give you an example of adaptation which raises even more awkward problems. I am thinking of the giant clams known as *Tridacna gigas*, which live off the Great Barrier Reef. (The name is a learned joke by the eighteenth-century naturalist Jean Guillaume Brugières: it means 'eaten in three bites', which certainly doesn't apply to giant clams as they can be as much as four feet long and two feet wide and may weigh 500 lbs.) They live in nutrient-poor waters and depend for their nourishment to a great extent on photosynthetic algae which live in their blood. What the algae get in return is unclear – it may be phosphorus and nitrogen. But how did the algae get in? The clams, which are perfectly equipped for digesting such plants, would promptly have digested them. Further, the clam has changed its shape completely, as compared with a cockle, twisting right round until the mouth tissues (which house the algae) and which would normally be underneath, are exposed to the sunlight. At the same time, they have generated a protective pigment to screen their tissues from the lethal effects of sunlight to which they are not adapted.

The world's leading authority on such creatures is Professor Sir Maurice Yonge. (He claims the distinction of being the only man to have read Gibbon's *Decline and Fall of the Roman Empire* from cover to cover while sitting on the Great Barrier Reef.) Baffled by the problem of how this amazing symbiosis evolved, he finds himself forced to the very un-Darwinian conclusion that the invading algae 'must have been fully adapted for life within animal tissues' before ever they got into the clams. But how the algae got transported from sunlight to shadow 'was a mystery then and remains a mystery now'.

While an adaptation of this kind is hard enough to explain in terms of natural selection, the case of our old friend the giraffe, which calls for a whole series of interlocked changes, is probably even tougher. No one gave much thought to the giraffe's problems until World War II, when the difficulties which pilots of fighter aircraft experience under severe accelerational forces caused biologists to look around to see how animals cope with a reduced blood supply to the brain. This led to the setting up of an international group for studying the giraffe after the war.

Nineteenth-century observers assumed that the giraffe had only to develop a longer neck and legs to be able to reach the leaves which other animals could not. But in fact such growth created severe problems. The giraffe has to pump blood up about eight feet to its head. The solution it reached was to have a heart which beats faster than average and a high blood pressure. When the giraffe puts his head down to drink, he suffers a rush of blood to the head, so a special pressure-reducing mechanism, the *rete mirabile*, had to be provided to deal with this. However, much more intractable are the problems of breathing through an eight-foot tube. If a man tried to do so, he would die – not from lack of oxygen so much as poisoning by his own carbon dioxide. For the tube would fill with his expired, deoxygenated breath, and he would keep reinhaling it.

Further, the study group found that the blood in a giraffe's legs would be under such pressure that it would force its way out of the capillaries. How was this being prevented? It turned out that the intercellular spaces are filled with fluid, also under pressure – which in turn necessitates the giraffe having a strong, impermeable skin. To all these changes one could add the need for new postural reflexes and for new strategies of escape from predators. It is evident that the giraffe's long neck necessitated not just one mutation but many – and these perfectly coordinated.

While some creatures evolve the most refined mechanisms in order to survive, strangely enough others seem to manage quite well with quite simple ones. One of the most adverse environments is the sand dune, which is very hot by day, extremely dry and constantly shifting. The few plants which manage to live in these conditions have a multitude of gimmicks to aid survival. Some turn their leaves edgeways to the sun. Some protect themselves with a feltwork of fibres, others by thickening the cutin layer which covers them. preventing water loss. Some reduce the number of pores in their leaves. But why don't all use all the tricks?

One of the more baffling features, in fact, is the many failures of adaptation exhibited by living things. Many plants have failed to adapt to the soils based on serpentine, just as they have not adapted to growth on mine-tailings. Many insects have not become resistant to pesticides, and birds show no signs of resisting PCBs and other environmental contaminants. The elm does not resist elm disease nor the American chestnut the chestnut blight. We ourselves have not adapted to the upright posture as the incidence of lumbar failure (slipped disc) attests. The external placing of the testicles is a major design error which evolution committed and seems unconcerned to rectify.

In short, the slick explanations of adaptation in terms of selection given in the textbooks ignore a lot of problems. The question which such texts evade is: is the wide range of variation found in nature due solely to natural selection or is it partly due to chance? And if so how are we to tell one from another?

In short, while some organisms prefer to adapt very little, others seem to revel in it. The reason is unknown and the facts fit ill with mutationism.

Sometimes the puzzle lies not in the variety of forms produced but in the almost wilful complexity of the adaptation, as if the designer had been a Rube Goldberg or a Heath Robinson rather than a blind machina or an omnipotent Creator. I am particularly entertained by the bizarre mechanism by means of which the cuckoo pint, *Arisaema*, which grows wild in my garden, achieves fertilisation. These plants have a conical 'spathe' surrounding and overhanging the spadix. When an insect enters the spathe of a male plant it is overcome by some unidentified vapour which the plant releases and falls to the bottom of the cone, becoming dusted with pollen en route. At the bottom, having recovered its wits, it escapes through a small hole provided as if on purpose. It then visits a female plant, is overcome as before, falls to the bottom depositing the pollen as it does so. But this time there is no hole! As it would be in the interest of the species to let the insect escape to fertilise other cuckoo pints, one can only conclude the absence of a hole is due to the cuckoo pint's black sense of humour.

Even if we concede the plausibility of the selective mechanism in principle, there remain a number of awkward facts which are regularly glossed over. For instance, some adaptations seem pointless or even harmful, yet they persist. Take the case of the sunfish, a creature fifteen feet long and weighing a ton and a half. In the course of evolution its spine has gradually reduced in length until it is now only half an inch long. No one can think of a good reason why.

Finally there are adaptations, useful at first, which progress too far, e.g. the huia bird of New Zealand. The male had a short stout beak with which it chiselled holes in trees containing the grubs of a beetle on which it fed. Unfortunately it couldn't extract the grubs, but the female had a thin beak twice as long as the male's which could get at the grubs. I can see that this gave the female an edge over the male but if either partner was carried off by some mishap, the remaining bird was liable to starve to death. Far from being an advantage in the struggle for survival such a specialisation seems most disadvantageous. One is not surprised to learn that the huia bird has become extinct.

Gavin de Beer dismissed this lightly as an instance of over-specialisation. So why did it occur? And how did these birds manage before they specialised? At what stage was there a selective advantage which would perpetuate the genetic change involved? Either the

The male huia bird had a strong beak needed for chiselling the bark of trees; but was dependent on the long beak of his mate for access to grubs. She in turn lacked a bill strong enough to make a hole. Each partner therefore depended completely on the other for food

two functions of chiselling and extracting were separated or they were not. De Beer's comment reveals a failure to consider seriously the very real problems for the theory of natural selection involved.

I have already, from time to time, mentioned several other instances of natural selection going astray: often this has been due to a trend going too far, as with the Irish elk and the tail of the lyre bird; at other times, it does not go far enough, as when a species fails to adapt to a change in climate or environment. Indeed, all extinctions of species and higher taxa (apart from major catastrophes) may represent a failure of adaptation. One is tempted to say evolution takes place in spite of natural selection rather than because of it.

Clearly many so-called adaptations are quite trivial, chance variations which crept in and proved neither advantageous nor disadvantageous. The problem which Darwin tackled, as we can now see, was not really the origin of species but the origin of adaptations. The origin of species, using the word in its strict genetic sense, is not too puzzling – except in the sense that we do not fully understand the mechanisms which deter one group of animals from interbreeding with its cousins. However, there is nothing inherently improbable about the phenomenon.

Another puzzling aspect of the question is why some creatures make an adaptation which seems, on the face of it, helpful, while other members of the genus manage perfectly well without it. For instance, one species of Peruvian armadillo, which lives high up, has evolved a fur coat, but other species living equally high have not. So where was the advantage?

Particularly difficult to accept as chance processes are those prolonged changes which lead to a new life-style, such as the evolution of birds from reptiles or – perhaps odder – the return of mammals to a life in the sea, as in the case of dolphins and whales. Even that fanatical defender of Darwinism Ernst Mayr says that these phenomena 'seem indeed as decisively directed towards perfect adaptation in the new medium . . . as if someone had directed the course of evolution'.

A few biologists have claimed that there is only one kind of adaptation, namely adaptation to a changing environment. It is not difficult to accept that if the climate is growing gradually colder, it would be advantageous to grow a coat of hair. Creatures that fail to adapt fast enough to such changes are liable to become extinct, they say. Leigh van Valen of Chicago University has called this the Red Queen hypothesis, an allusion to the lady in *Alice Through the Looking Glass* who had to run as fast as she could to stay in the same place. Van Valen supports his case by showing that the rate at which species become extinct is much the same, regardless of the length of time for which they have existed. If natural selection were actually improving the fit of organisms to their environments, he argues, the species which have existed longest might be expected to have a lower rate of extinction, since presumably they would be better adapted to their niches.

While failure to adjust to environmental changes may well be a major factor in extinction, such adaptation is certainly not the only kind of adaptation. In a few cases, the answer seems to be: a happy chance. For instance, in the weasel, a slip of muscle is detached from the main shoulder muscle. This makes no difference to the weasel, but when the same feature appears in the badger it greatly improves its powers of digging with its fore-paws. In the same way, hydrocyanic acid doubtless first appeared in clover by chance. The clover plant has evolved the power of producing this highly poisonous substance, which discourages some herbivores from eating it. A good many other plants have done the same; and some herbivores have learned how to cope with this by detoxifying what they eat. (This ploy has the disadvantage that, in very cold weather, the cells containing the hydrocyanic acid are liable to burst, thus poisoning the plant itself. Thus a modification which favours survival in one environment is adverse to it in another.)

But the main driving force in adaptation is the animal's behaviour. It is more likely to experience a changed environment because it has moved to a new location than because the environment itself has changed. When bears moved to the Arctic, they adapted by becoming white. The flattened face of the Eskimo and his squat body are said to be an adaptation to living in cold conditions, just as the black skin of Africans is said to be an adaptation to strong sunlight. So the question arises, why do creatures move out of niches to which they are adapted and into ones to which they are not adapted – and then have to wait for many generations for natural selection to do something about it? On Darwinian arguments, this would be a suicidal policy.

Certainly Darwin does not emerge unscathed. If it is true that speciation occurs mainly where there are plenty of open niches, then it is clear that competition is adverse to speciation, whereas Darwin makes competition the driving force of selection. Odder still, the record suggests that newly evolved organisms wait for the old to vacate their niches before occupying them, even though they are more efficient. Thus the rodents are generally thought to have been a step ahead of the multituberculates (an early order of mammals with rodent-type teeth) and gradually replaced them. Crocodiles replaced the rather similar phytosaurs. Why are there no longer any forms intermediate between herbivores and carnivores?

Again, within one square metre of ground a score of species of snail may be found. What advantage can any one of them have? Persistence of unneeded characters is hard to explain.

5 *Evolution Upside Down*

Now we can begin to draw the threads of the discussion together and ask three questions. (1) Does the process of natural selection explain speciation adequately? (2) Does speciation account for the much larger differences between orders and phyla, as Darwin claimed? And (3) how far is the mechanism of evolution based on chance?

There seems no need to doubt the basic premise that straightforward modifications, such as the length of a bird's beak, the colouration of a moth or the length of a wading bird's legs have been fostered by selection operating on a genetic change of some kind. However, it seems to me that certain qualifications need to be made, qualifications which you will not find in the textbooks. First, adaptability of different organisms varies and so does their ability to respond to particular pressures. The primrose, for instance, responds with a much narrower range of modifications than the orchid. An organism which can adapt to extreme heat may be unable to cope with damp conditions. It is a reasonable guess that such adaptabilities may have varied over evolutionary time, which would go far to explain the explosive radiations to which I referred. (In the next chapter we shall see what genetic mechanisms might bring this about.)

Second, some modifications are probably neutral, neither selected for nor against. Many of the variations in the form of orchids can have little selective value; or, at least, one variant is not more advantageous than another. It matters little whether a flower looks like a clown or a tragedian provided it attracts the fertilising insect.

Third, I cannot doubt that trends become established because they are of benefit initially and then continue far past the optimum. A development of the incisors to the point where you cannot seize anything in your mouth is clearly anything but advantageous. The mechanism behind such 'overshoot' is not understood.

Perhaps the most pregnant of the questions associated with speciation is whether it truly accounts for the major divisions into phyla which have remained stable for so long. The German thunderer, Richard Goldschmidt, is one who thought not. He says that evolution proceeded 'from the top down', from the phylum to the species.

Commenting on the fact that a phylum contains classes, and so on, he writes: 'Can this mean anything but that the type of the phylum was evolved first and later separated into the types of classes, then into orders, and so on down the line? This natural, naive interpretation of the existing hierarchy of forms actually agrees with the historical facts furnished in paleontology. The phyla existing today can be followed furthest back into remote geological time. Classes are a little younger, still younger are the orders, and so on until we come to the recent species which appear only in the latest geological epochs. Thus logic as well as historical fact tell us that the big categories exist first, and that in time they split in the form of the genealogical tree into lower and still lower categories.'

To this Mayr makes the remarkably feeble reply that the categories are 'a man-made artifact'. If I assert that the partition walls of a house were put in after the main structure had been erected, this is in most cases quite unaffected by any difficulties in defining the word 'house' or the word 'wall'. I doubt if such a position can be sustained. I think rather that the answer is to be found in genetics – that, long ago, the genetic material took some irreversible steps, which produced certain limits to variability. Within these limits there would be further decision points, and so on. Here, as elsewhere, we have to turn to the genetic control system for an answer and I shall pursue the point in the next chapter.

On one thing even the most stubborn Darwinians are agreed; the real regulator of speciation is the behaviour of the creature concerned. Ernst Mayr concedes: 'A shift into a new niche or adaptive zone requires, almost without exception, a change in behaviour . . . to be followed by a change in structure.' Changes of habitat, changes of feeding habit, changes of mating preference: these and many other behavioral features are what open up new niches. But this is as much as to say that evolutionists are studying the wrong subject. Unless we can discover why one behaviour is displayed rather than another we cannot pretend to say why evolution went the way it did.

Though evolutionists often pay lip service to the importance of behaviour they then ignore it. I shall therefore return to this all-important topic in the penultimate chapter.

I have kept up my sleeve one final, crushing proof that we still do not understand speciation. The most prominent thing about evolution is that it is going somewhere. Organisms are becoming more complicated, their capacities are becoming more sophisticated. Species formation absolutely fails to explain why this should be. Behaviour too becomes more sophisticated. Why should this be?

The great statistician who devised population genetics, Sir Ronald Fisher, once said that evolution was a device for generating improbability. That about sums it up.

CHAPTER EIGHT

Revolution in Genetics

1 Passing the Buck

Fundamentally, the evolutionists resolve all their problems by pushing them over to the geneticists and assuming that they are capable of dealing with them. That is to say, for every variation which appears in nature, the evolutionists assume the cause is a mutation or genetic change: they do not enquire whether the genetic system is in fact capable of producing the desired structural changes and doing so in the available time. In fact, the capacities and limitations of the genetic system are turning out to be quite other than they supposed.

With all that we now know, it is difficult to comprehend the confusion of ideas which existed fifty years ago, a confusion which neo-Darwinian theory seemed to clear up, much to everyone's relief. In the 'twenties and 'thirties many biologists did not believe in the theory of natural selection at all, especially in France, where Lamarckism ruled supreme. There were at least five theories of evolution under discussion. Geneticists and paleontologists never met. They read different papers, went to different meetings. An attempt to bring them together at a conference in the 'twenties ended in confusion and misunderstanding.

Genetics itself was split down the middle. There were those who, as followers of Mendel, believed that evolution was gradual (as Darwin had said) and those who, impressed by the abrupt mutations which de Vries thought he had discovered in the evening primrose, were convinced that evolution was jerky. When, in the 'twenties, H.J. Muller showed that there were indeed mutations – by bombarding fruit-flies with X-rays – they felt confirmed in their belief. No love was lost between the two factions. Mayr, recalling those days recently, said: 'When I read what was written by both sides during the 1920s, I am appalled at the misunderstanding, the hostility and the intolerance of the opponents. Both sides display a feeling of superiority over their opponents "who simply do not understand what the issues are."'

When the Mendelians peered down their microscopes at dividing cells, they saw the chromosomes splitting and pairing, and noted how sometimes sections broke away and were inserted elsewhere, sometimes in another chromosome and perhaps inverted as well. They fervently felt that these 'rearrangements' were sufficient to account for evolution. Meanwhile their rivals who by now had huge populations of fruit-flies which they were subjecting to stress from heat, chemicals and anything else they could think of, were beginning to map on the chromosomes the actual positions of some of the mutated genes.

The naive belief of both parties that genetics alone could account for evolutionary change was based on a grossly over-simplified notion of the nature of the genome and genetic mechanism. For long it was believed that the genes were strung together 'like a string of beads' and that each gene controlled the expression of a single character. There would be one gene for eye colour, so that a mutation of a single gene could change eye colour; another would cause polydactyly, and so on. And when, just after the war, Caltech's George Beadle, working with bread moulds, showed that each gene produced one, and only one, enzyme, this seemed to confirm the idea. However it gradually emerged that most characters, even simple ones, are regulated by many genes: for instance, fourteen genes affect eye colour in *Drosophila*. (Not only that. The mutation which suppresses 'purple eye' enhances 'hairy wing', for instance. The mechanism is not understood.) Worse still, a single gene may influence several different characters. This was particularly bad news for the selectionists, of course.

However these complexities were as nothing to what was to come. One of the lesser puzzles which must have irked those evolutionists who knew about it was the fact that some characters exist in more than one form: the best-known example being the four human blood groups. (This is known as polymorphism.) Why has natural selection not eliminated all but the most efficient of these blood types or why have genes for malaria resistance not become general in the population? Until recently it was assumed that such polymorphisms were so rare that they could be neglected, and the matter was brushed under the carpet. But in 1966 Henry Harris of London University demonstrated, to everyone's surprise, that as much as 30 per cent of all characters are polymorphic. It seemed unbelievable, but his work was soon confirmed by Richard Lewontin and others. Recovering, the geneticists hastily pointed out the potential advantages of polymorphism – a supply of alternatives ready to meet altered conditions and so on. But how is polymorphism maintained in the face of natural selection? There must be special mechanisms to defeat it. 'If this conclusion is correct,' says Bryan Clarke of Nottingham University, who has made an extensive study of polymorphism in snails, 'it has several important implications. To begin with, it requires that we revise our ideas of how evolution proceeds.'

Then as now genetics was unable to answer the main question: can chance account for the appearance of a wanted or useful mutation of form at the right moment, and do it again and again? This was only to be expected. But it was also unable to demonstrate a mechanism by which, for instance, the woodpecker could acquire a reversed claw. Much less could it explain the processes by which an intricate organ like the ear was formed. The position is no more encouraging today. Indeed, the discoveries of the 'sixties and 'seventies have made it difficult to account for even the simplest modifications. The genome has turned out to be so intricate, so much is going on within it which had never been suspected, that if we succeed in grasping the principles at work, genetics will be lifted on to a new plane and all may become clear. In what follows I shall try to give you an idea of the radical discoveries made in the last few decades.

2 *DNA and the Frame Shift*

Until 1944, for some sixty or seventy years, it had been confidently assumed that the genetic material was composed of protein. The chromosomes were known to contain both protein and DNA, but the latter was an obscure substance about which little was known, while evidence about proteins was accumulating rapidly. As you may recall, protein is the name of a very large family of substances whose common feature is that they are composed of strings of amino-acids. There are some twenty amino-acids and the nature of the protein is determined by the precise sequence in which these occur in the molecule. Muscle is the best-known example. The chains are very long, consisting of hundreds of thousands of amino-acids, and they fold themselves automatically into strange configurations, thanks to various forces which act between different amino-acids. Proteins are synthesised from amino-acids by all living things.

It was therefore a bombshell when Oswald Avery and colleagues at the Rockefeller Institute demonstrated in 1944 that it was the DNA which carried the genetic message. By transferring the DNA from a rough-coated bacterium to a smooth-coated one, he caused the latter to produce rough-coated offspring, and the character bred true thereafter. It was an epoch-making experiment and why Avery did not receive a Nobel award for it is one of life's great mysteries.

In the years following, several experiments confirmed and extended this finding, and interest focused more and more on DNA. How was it composed? How did it encode a genetic message? How was anything so complex accurately duplicated at the time of cell division? The breakthrough year was 1954 when Maurice Wilkins, using a technique borrowed from physics, showed that it has a double helical structure, while Francis Crick and James Watson proposed a model which would account both for the coding and the duplication. The story has been told many times but I will recapitulate it, since an understanding of the mechanism, as then understood, is essential to the appreciation of the still more complex picture which is emerging now.

The chemists showed that DNA is composed of a string of units called nucleotides, which come in four varieties, the central part having one of four groups (known as bases) attached to it. They are conveniently known as A, C, G, and T, from the initial letters of the substances composing the bases. Now these bases have the extraordinary property that A and T readily cling together, and so do C and G. Thus a single DNA chain, immersed in a pool of nucleotides, will promptly pick up a T for every exposed A, a G for every exposed C and so on, thus forming a counterpart chain. This at once solved the problem of duplication. For if, at cell division, the double chain is opened up and pulled apart, each half will rapidly create its own counterpart.

Where Crick showed his greatest originality, however, was in proposing that the DNA chain should be read in groups of three. For instance, where the DNA chain contained, consecutively, CAT the amino-acid valine will be incorporated in the protein, CCT will lead to the incorporation of glycine, and so on. In short there was a code, the details of which took some time to unravel.

The machinery whereby amino-acids are linked together in the order thus prescribed need not really detain us, except that it was quite unlike any chemical process previously known. The DNA message is copied onto a single strand of RNA, a substance almost identical with DNA. RNA performs several functions: lengths of it, generated by DNA, curl up into little machine tools known as ribosomes. Other bits of RNA attach themselves to free amino-acids and present them to the ribosomes. The RNA tape runs through the ribosome like a tape through a tape-controlled machine tool, specifying which amino-acid is to be incorporated next. It is all more like a factory than a chemical reaction.

The effect of this cluster of suggestions, all borne out by the facts, was stunning. Nothing of this kind had been conceived before. In the excitement, the extreme originality of Crick's thinking was perhaps not fully realised. He had, first, to get rid of the idea that proteins were the determining items. Then he had to envisage that transcription of the DNA took place sequentially. People were used to thinking in terms of assembling units on jigs and removing them en bloc when ready. The idea of a process which could go rolling on and on indefinitely was new. Finally, the idea of a code, in which several things would specify one thing, was quite new in biology. It enabled DNA with its mere four units to specify the composition of proteins with their twenty components.

Incidentally, the four bases, A, C, G, and T can be arranged in many more than twenty ways: in sixty-four to be exact. And in fact some amino-acids are specified by several different combinations. Other combinations act as stop-and-start signals.

The DNA of a typical bacterium comprises some three million nucleotides coding for some 3,000 genes. The DNA of man contains some three billion nucleotides, but only codes for – at most – 150,000 genes and possibly as few as 30,000.

To sum up the effect of these discoveries: the gene which was once thought to be a discrete unit, either present or absent, functioning or suppressed, now turns out to be a chain of a thousand or so nucleotides, any one of which might be deleted or damaged, thus impairing or making nonsense of the message – i.e. failing to construct the required enzyme. Moreover, to bring about a meaningful alteration, such as the production of a different enzyme, would mean the altering of various individual bases among the thousand or so in the gene in a precise manner. How such alterations could come about by chance is wholly incredible.

There is another curious consequence of the coding system which may be of great importance: errors in transcription can arise by taking the nucleotide triplets in the wrong groups of three – so-called 'reading frame' errors.

Suppose we have a string of nucleotides like this: AAA, ACG, AAA, CCG, AAG, CAC, CCT. They will select the following amino-acids: lysine, threonine, lysine, proline, lysine, histidine, proline. Now let us suppose that by some accident – the impact of a packet of ionising radiation, for instance – the A is deleted from the second of the above triplets. The series of nucleotides will now fall into a different lot of triplets, as follows:

AAA, CGA, AAC, CGA, AGC, ATC, CT . . .

and these will, of course, specify a different sequence of amino-acids, thus:

lysine, arginine, aspartic, arginine, serine, isoleucine . . .

Thus an error affecting less than one-hundredth of the gene's components could produce a wholly different product. Furthermore, if a section of the chain is transposed to another part of the chain and the break occurs in the middle of a triplet, similar but even more confounded confusion will result. Finally, damage to a single nucleotide could alter stop-and-start signals. Obviously end-for-end inversion would have even more drastic effects.

If this is confusing, we can think of an analogy: omissions and transpositions in the words which compose a sentence. (Here our code is written in twenty-six letters, not just the four letters to which DNA is limited.) Thus the sentence THE CAT SAT ON THE MAT could become THE RAT SAT ON THE HAT by two substitutions. An insertion might make THE CAT SPAT ON THE MAT. A deletion: HE-CAT SAT ON THE MAT. In short, there are many more possibilities of mutation than were realised, and the problem is now seen to be not, where does the variability come from but how is a reasonable degree of stability maintained? True, the great majority of mutations (using the word to mean damage to the DNA system) will be harmful or lethal. The code will become nonsense. Only if the change happens to produce an enzyme which is advantageous will it be preserved, and in a moment I shall explain how such fortunate mutations are quickly exploited and amplified.

There are, it is now established, specific chemical substances which have specific effects on the process of DNA replication. For instance, the acridine orange group of dyes causes deletions, and insertions of single nucleotides bringing about a 'frame shift' of the kind just described. Alkylating agents like mustard gas seem to scoop G (guanine) out of the chain, allowing any other of the three units to substitute, others cause A to behave like T, and so on.

Meanwhile work was going on which would put all these discoveries in the shade, while giving more precision to the notion of a gene.

3 *The Protean Genome*

By 1960 molecular biologists were happily pursuing the host of problems evoked by the elegant new theory and the weather seemed set fair. In particular they wanted to establish the actual sequences of nucleotides in specific genes and the actual sequences of amino-acids in specific enzymes. At that time, the task of identifying in turn each of a thousand or so chemical units was incredibly laborious. It could take a whole team five years to unravel one. But gradually methods were improved and, to some extent, automated so that now sequences are streaming in and it has become necessary to establish a central reference library to prevent anybody doing the same job twice. Progress was rapid.

In 1969 Jonathan Beckwith's team at Harvard Medical School isolated a complete gene from a bacterium. The following year H. Gobind Khorana at the University of Wisconsin, with the support of a large group, actually synthesised a complete structural gene, an historic achievement. The information he needed however had been assembled in 1964 by Robert W. Holley who determined the nucleotide sequence in the corresponding RNA. Both shared a Nobel prize later. It was a landmark in the history of biology. Soon the synthesis of other genes was being attempted and it was being demonstrated that these artificial genes would do their jobs in the living cell as well as the natural matter.

But in the 'sixties cracks began to appear in the intellectual structure, culminating in discoveries so shattering as to leave the whole edifice tottering. Reports began to come in of new genes appearing 'from nowhere' and of groups of genes mutating simultaneously.

As Francis Crick wrote in *Science* in 1979, 'In the last two years there has been a mini-revolution in molecular genetics.' He was thinking mainly of the discovery, by Pierre Chambon of the Louis Pasteur University in Strasburg, of the startling fact that genes need not consist of a continuous sequence of nucleotides but may be split into sections separated by 'nonsense DNA'. The nonsense sections are now known as introns, the gene-forming segments as exons.

When he made his discovery, Chambon was actually investigating how the hen manufactures albumen for its eggs, a process controlled by a single gene. He set out to clone the gene. To his astonishment, he found that this gene was composed of eight exons, each with an intron. Work in other laboratories soon confirmed this, and it became clear that the whole shlamozzle, introns and exons, are transcribed into RNA, whereupon a splicing mechanism, still being explored, cuts out the nonsense, joins the effective fragments and passes them to the cytoplasm (the part of the cell outside the nucleus) to manufacture protein.

These split genes have now been found in most higher organisms: mammals, birds, amphibians, insects and plants.

Geneticists were quick to see that this queer arrangement had evolutionary advantages. New genes could arise by fusing the genetic units, without losing the old ones. Evolution can seek new solutions without losing the old. On this view, comments Walter

Gilbert, introns 'are frozen remnants of history and sites of future evolution'.

However, Chambon thinks there may be more surprises to come. 'The possibility cannot be excluded,' he says, 'that introns are implicated in very different functions still to be discovered . . .' – functions affecting the organisation and expression of genetic information.

Despite the potential advantage, I find something distinctly implausible about such an arrangement. We are quite in the dark as to how the RNA cuts out the nonsense. However, an even more implausible mechanism was already known. In the late 'forties, Barbara McClintock, at the Carnegie Institution's Department of Genetics, had been studying the pigmentation of maize. She was puzzled by the fact that certain genes were being turned on at abnormal times, as shown by the appearance of pigmented grains among the normal ones. But this didn't make sense by the criteria then current and her observations were put in the 'funny results' bin and forgotten. Twenty years later, a group at Tufts University, with James Shapiro, then at Cambridge University, found a similar phenomenon in the intestinal bacterium, *E. coli*.

They found a new type of mutation. It emerged that certain genes were capable of jumping from their normal position in the DNA chain and inserting themselves elsewhere. They carried with them a special enzyme which opened the chain and resealed the ends. When this happened, the gene which they invaded was incapacitated and very often the next gene in sequence was also affected. These 'jumping genes' were capable of jumping to many different sites but certain sites seemed especially favoured and became known as 'hot spots'. Lengths of between 2,000 and 4,800 odd nucleotides were involved – enough, that is, to comprise two to four genes.

Bacterial cells contain small rings of DNA known as plasmids; they are virtually rogue genes with their two ends hooked together. It emerged that some of these, at least, could spring into activity and insert themselves in the main DNA chain.

A British team discovered that resistance to antibiotics could be transferred from one plasmid to another, and suggested the name 'transposon' for what had previously been hesitantly called Inserted Sequences. It emerged that transposons could also delete other genes, or promote their action, or invert lengths of DNA.

As if this was not enough, Austin Taylor of the University of Colorado discovered in 1963 a peculiar bacterial virus which could insert its DNA into bacterial DNA at many different sites, causing many different kinds of mutation. (Ordinary bacterial viruses can insert their DNA but only at one site.) He called it Mu, for 'mutator'.

Further studies suggest that it is basically a transposon which can also exist as a virus.

All this work was done with bacteria. It remains uncertain how far the findings apply to higher organisms, in which the DNA is more closely protected, but since transposons have been found in yeast and fruit-flies, as well as plants, it probably does.

All this put the cat among the pigeons as far as genetics was concerned. From a situation in which spontaneous mutations were held to be too few to account for evolution, they had now become almost too plentiful, as well as too erratic. Variation is generated, it seems, not by mutations but by the shuffling of large pools of mutations which have been accumulated over many generations. As Stanley N. Cohen, one of the leaders in the field, commented in some surprise: 'The discovery of such a fundamentally different recombinatorial process, at a time when many molecular biologists believed virtually all the important aspects of bacterial genetics were understood in principle . . . leads one to wonder whether still other fundamental new and significant basic biological processes remain to be discovered.'

For evolutionists, naturally, all this was of prime importance, for if DNA segments which have little if any ancestral relationship can be conjoined, this would cause quantum jumps in evolution.

4 Genes Within Genes

To cap all this came a discovery which I find quite incredible and which carries the notion of gene fluidity still further. A small bacterial virus known as phage 174X had long puzzled molecular biologists because it did not seem to contain enough DNA to specify the nine proteins of which it is composed. The puzzle was solved when Fred Sanger, who won two Nobel awards, one for devising a method of sequencing proteins, the other sequencing DNA, applied his technique to it. He found that the base-sequence specifying two of the proteins was, so to say, embedded in the genes for two other proteins, and were elicited by a reading-frame shift of the kind I have just discussed. Not only that, but one section of DNA is actually read *three* ways. That genes can exist within genes obviously increases vastly the number of genes a given length of DNA can embody and makes calculations of the total number of genes in an organism on the basis of the weight of its DNA difficult or impossible. As I shall explain in the last chapter there are great difficulties in seeing how such an embedding of genes within genes is possible. Here again it is still uncertain how far the same thing occurs in higher organisms as occurs in the tiny bacteriophage.

This finding also reveals the gene as subtler and more elusive than anyone had supposed.

From the time of Crick's and Watson's model it had been a matter of dogma that the DNA was extremely stable. While information could get out of it, it could not get in – so most biologists supposed. For if DNA kept altering, how could we account for the relative stability of the various living species? Hipparion may evolve into the modern horse, but it does not suddenly grow horns or turn into a rhinoceros. It was clear that the genetic code itself had long remained unchanged since the beginning of the story, for DNA from advanced species, inserted into cells from primitive forms, continued to work normally. DNA can even be transferred from animals to plants in some circumstances and carry out its duties. Sequencing of DNA from more than sixty animals of varying degrees of evolvedness has recently confirmed how little change has occurred. But though the language may not change, the messages conveyed by it may. The internal instability of the genome represents a major unsolved problem for evolutionists as well as geneticists. What a long way we have come from the earlier notion of the gene as a bead on a string!

There was yet one more unexpected discovery of the startling 'sixties: in 1964 it was found that most (perhaps all) cells contain multiple copies of some of their genes, and of their non-gene (plasmid) DNA also for that matter. Since all the evidence was for a very precise mechanism, this loose arrangement came as a major surprise. The discovery was made by a group at the Carnegie Terrestrial Magnetism Laboratory who found that certain experiments which detect small amounts of DNA worked much too well. Mouse DNA sequences showed up in them much quicker than bacterial DNA, contrary to expectation.

The amount of this 'redundant' DNA varies from 20 per cent to as much as 80 per cent, according to the species. It is known that these copies are often imperfect, but whether they constitute regulatory DNA or structural DNA is uncertain. It seems, too, that only certain genes are copied. Thus in one study of calf thymus it was found that there were two families of repeated sequences, one present in a million copies, while the other was present in 66,000 copies. How did it originate? Does it have a function? These are unanswered questions.

More recently, the idea has been advanced that such duplicate copies may enable the cell to manufacture large amounts of some suddenly needed substance by producing the needed enzymes simultaneously. This seems reasonable where the number of copies is a few hundreds, as sometimes occurs, but is implausible where a million copies are involved. The fact that the copies are not identical also seems inconsistent. Biologists now talk of 'gene amplification' and 'multi-gene families'. Recent work suggests that these copies are unstable, constantly breaking down and being renewed.

It seems to be the case that amplified DNA is used to produce the ribosomes or 'machine tools' when these are wanted in a hurry, and some have wondered if the phenomenon could be connected with the rapid cell multiplication in cancer. One of the great unsolved mysteries is whence, in the course of evolution, did new genes come? It would be absurd to suppose that the primitive amoeba or bacterium carried genes for wings, haemoglobin, rhinoceros horns and all the myriad structures which were later to manifest. So the original genome must have been progressively added to. (Incidentally, this is a conclusive argument against those who maintain that the genome cannot be altered. It can certainly be added to and probably genes can be lost too.)

It might reasonably be thought that the amount of DNA in the genome would increase pretty steadily as we advance up the evolutionary scale, for this reason. But in fact measurements of total DNA content are quite confusing. While the mammalian cell seems to have about 800 times more DNA than a bacterium, toads (to take an example) have very much more than mammals, including man, while the organism with most DNA (of those so far studied) is the lily, which can have from 10,000 to 100,000 times as much DNA as a bacterium!

The most daring hypothesis, however, is that this redundant DNA might comprise a store of mutant forms which the cell is waiting to try out or just to get rid of. In any case it must surely be the raw material out of which new genes are to be manufactured. The suggestion poses radical problems for we know of no way this could be done, but new genes *do* arise so there must be mechanisms we know not of.

It will now be very clear that genetics has turned out to be something very different from what the evolutionists assumed – and no doubt there are further surprises in store. Francis Crick, commenting on the discovery of gene splicing, wrote in 1979: 'It is virtually certain that discoveries will turn up which will radically alter our ideas of the details of the evolutionary process.'

It may be that it is not genes (that is, strings of nucleotides) which determine structure so much as the relationships between such entities. If the basic assumption that evolutionary variety depends on chance rearrangements of such entities is proved false then the lynch pin of evolutionary theory as it has been known in this century is gone.

CHAPTER NINE

Transformation Scene

1 Regulation and Control

When the spring comes, the trees are dusted with green and leaves begin to form. Flowering plants not only form new shoots but also begin to construct the elaborate, multi-coloured structures we call flowers. What has happened at the genetic level? Clearly a messenger has arrived in the appropriate cells and has said: 'Leaf genes and flower genes, get going. Do your stuff.' These genes, which have been switched off all winter, are now abruptly switched on. In the language of genetics, they have been repressed and are now derepressed.

Genes, as this example shows, do not work all the time. Gene repression is widespread; it ranges from the genes which form secondary sexual organs, which are silent from birth until puberty, to genes which control the manufacture of digestive juices, and are derepressed two or three times a day. Other genes are only switched on when need arises, such as those which promote the regrowth of skin and tissue when a wound is healing.

If I labour this obvious fact it is because many evolutionists have overlooked it. They often write of a mutation occurring and either being selected against or favoured by selection. But a mutant gene will only be exposed to selection if and when it is expressed. Hence the question of how and when genes are expressed is central to the problem of evolution.

In point of fact, the control of gene expression has become one of the leading areas of interest in genetics. The subject has been transformed by the discovery that the great majority of genes are regulatory. Clearly, a mutation in a regulatory gene could bring about a substantial change in the appearance of an organism at one blow. No longer is there any need to wait for the simultaneous appearance of several different structural mutations.

There would seem to be three time spans over which regulation occurs and hence perhaps three different mechanisms to control expression. First, there is the moment-to-moment adjustment of, say, the output of a hormone. In the years just after the war Jacques Monod and Jacob Wolman showed how this is achieved and won a Nobel award for their achievement. Second, there are controls which only come into action occasionally, such as those which give the snow hare its winter coat or function only in the mating season. Lastly, there are decisive, long-term controls. We see these at work in the development of the embryo. For instance, a section of tissue destined to become skin will, if transplanted to the mouth area at an early stage, become part of the mouth. Genes which would have formed skin have been switched off and mouth-part genes switched on. At a later stage, the presumptive skin is unable to respond to the local influences and remains skin. Its skin genes have been irrevocably switched on. In the 'sixties, F. C. Steward, Director of the Laboratory for Cell Physiology, Growth and Development at Cornell, stunned the scientific world by taking cells from a carrot root and bathing them in coconut milk; some of them began to develop into carrot plants with normal roots, stalks, flowers and seed. In this process (now famous as cloning) the whole array of genes necessary to produce a mature plant was switched on again in cells which had already specialised.

Undoubtedly, these regulator genes must be arranged hierarchically. Thus there will be a group which organises the finger and contains within it genes which organise nails, skin, muscles and tendons. But this gene will be a member of a larger group which forms the hand, and this in turn part of a gene array responsible for the whole arm. Thus mutational damage to a high-level gene could cause failure of a whole organ to form, or the formation of the wrong organ. We can in fact see this happening in organisms like the fruit-fly, where traumatic treatment can cause an eye to form where a haltere (balancing organ) was expected. More often a wing substitutes for the haltere, which is itself a highly modified wing.

If something of this sort occurs, it becomes easier to understand the abruptness of the changes in Eldredge's trilobites, or even the suddenness of the major evolutionary changes we discussed in Chapter Three.

There is a special case of whole blocks of genes being switched on and off which, though studied in its own right, has not been given much consideration in the context we are now discussing, namely metamorphosis – the transformation of tadpole to frog, of chrysalis to butterfly, of larva to jellyfish. How many of those who have gazed, as a child, at a butterfly emerging have stopped to wonder how such a thing is possible. Let us give the matter our attention.

2 *Two for the Price of One*

Of all the varied phenomena of development, surely none is more extraordinary than metamorphosis. Most of us, as children, have collected tadpoles from a pond and have watched them produce legs and then suddenly turn into a frog. Within a week or so, an organism adapted for living under water on vegetable fragments and swimming by means of a powerful tail has become transformed into a four-legged animal living on land and consuming insects. In addition to the obvious changes of shape there are less obvious biochemical changes. The haemoglobin of a frog is different from that of a tadpole. Different digestive enzymes are required. Fish excrete ammonia, which is at once diluted by the surrounding water. Land animals cannot afford to do this or they would be poisoned, so they convert the ammonia to harmless urea. So the frog switches to producing urea where the tadpole produced ammonia. (Some fish have also learned this trick.) Like other land animals, the frog conserves water by means of a thick, impermeable skin, whereas the tadpole has a thin, permeable skin.

These changes are exquisitely timed. The tail does not begin to vanish until the legs are ready. The gills do not resorb until the air-pumping mechanism is functional. In the metamorphosis of insects the change is even more striking. A larva is singularly unlike a butterfly both in appearance and mode of life. The structural changes are far reaching, involving nerves, muscles, gut, sense organs, respiratory system, circulation, fat bodies, skin, alimentary canal and so on. As John Whitten of Northwestern University has said: 'In fact there would appear . . . to be no system that remains unaffected by the metamorphic changes occurring during the pupal stages' in those insects which display metamorphosis in its most developed form.

Apart from the many difficulties in understanding how such a radical change comes about, there is the larger question of why it should happen? Can there really be an evolutionary advantage in constructing one sort of organism and then throwing it away and starting again?

Metamorphosis, it must be conceded, does not always take place with such dramatic suddenness as in the frog or the butterfly. The salamander before metamorphosis looks pretty much like the sala-

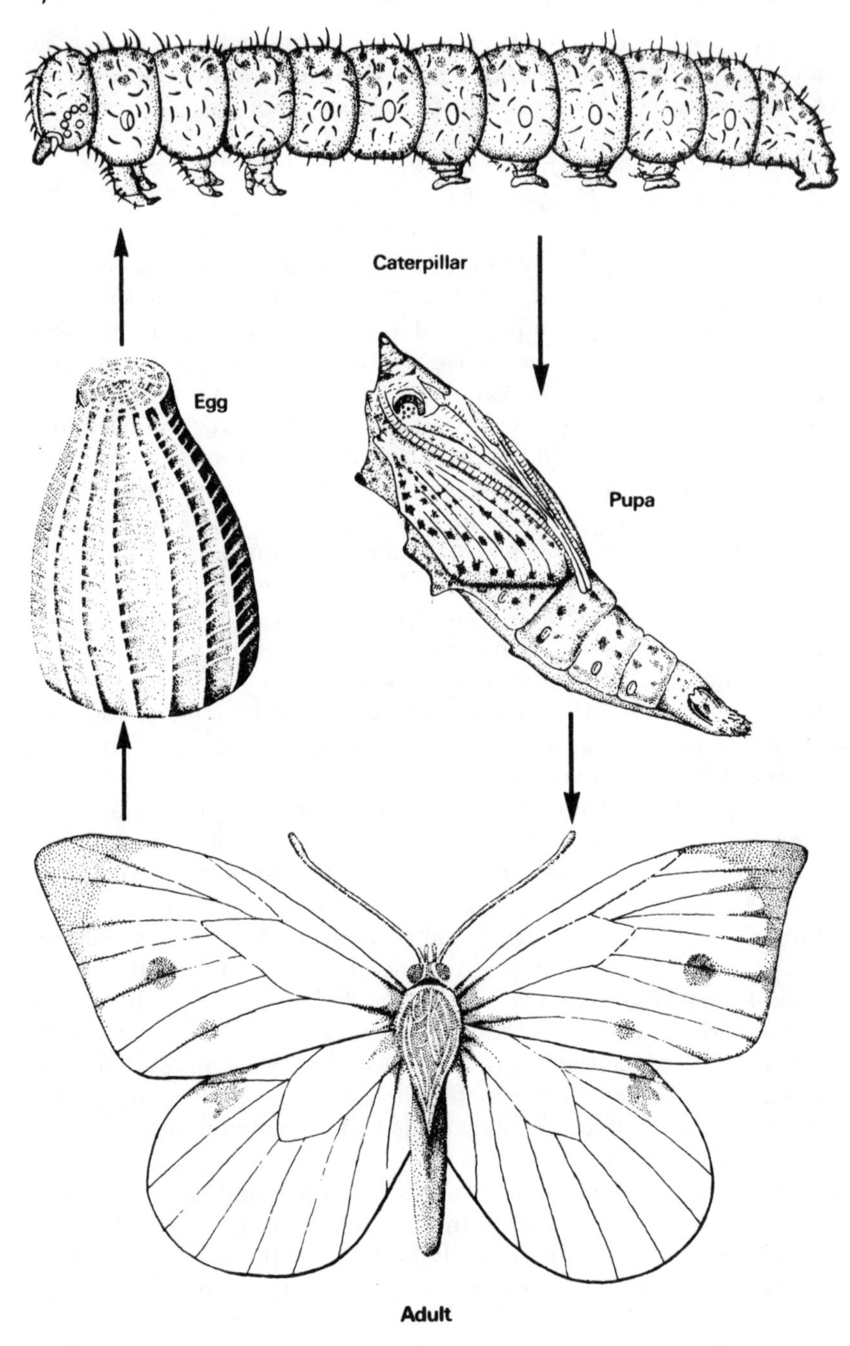

Stages in the development of the individual butterfly, starting from an egg here scaled up for clarification

mander after metamorphosis: both versions have legs and a tail, the tail is somewhat reduced, the legs rather more developed after metamorphosis, that is all one would see at a glance.

The life of the pre-metamorphic form can vary widely. Whereas the tadpole in the smaller species exists only for two or three weeks, those of the bull-frog flourish for three years before metamorphosing. In some crustacean metamorphoses the larval forms persist only for a few minutes. In shrimps the change is not sudden but is spread out over a long period. So many variations are found, ranging from simple adaptations to complete reorganisation, that John Costlow of Duke University regards the phenomenon as one which evolved separately in each order.

In insects, with their successive moults and pupal stage, the process is particularly closely integrated. Those parts which will not later change in shape harden before pupation; those, like the wings, which require to expand, harden after emergence. The pupal case itself is shaped to accommodate the butterfly that will be rather than the larva which is, a subtle piece of planning which can hardly be due to chance.

A particularly puzzling feature of metamorphosis is what happens to the structures which are resorbed, notably the tail. Resorption starts with stunning rapidity: changes in the tail can be detected within two or three minutes of onset of the metamorphic process. What happens to the material? Is it re-used? And what destroys the cells – an agent from within or an agent from without? The whole process is a nest of puzzles.

I am particularly struck by the case of the red eft. There is a variety of newt known as *Notophthalmus viridescens* which, after an aquatic phase, metamorphoses into a creature popularly known as the red eft, which spends from one to three years on land. It loses its lateral line (which, you will recall, is an organ peculiar to fish) and gains the kind of tongue which is useful in catching insects. Then, after anything up to three years on land, it metamorphoses back again, losing its tongue and regaining its lateral line. It is evident that the genes for forming the lateral line were not lost, only suppressed. To confuse the situation further, I must add that the clawed toad, *Xenopus laevis*, a darling of biologists, never leaves the water. Both larva and adult are tongueless and have a lateral line. That fits with their continued aquatic life, but leaves one wondering why they bother to metamorphose at all. As two British scientists, Baldwin and Underby, showed, *Xenopus* makes a half-hearted attempt to shift from ammonia excretion to urea excretion, as frogs do but then gives up the effort.

One thing can be confidently deduced. There are clearly two

distinct sets of genes involved: in the case of Anurans a set that specifies tadpole and a set that specifies frog. At metamorphosis, the tadpole set becomes masked or switched off, the frog set becomes unmasked or activated.

Recent work on insects suggests that the distinction may not be so absolute as this suggests. It is not difficult to understand that a gene which causes the formation of muscle in a tadpole might also cause formation of muscles in a frog, even though the muscles might be different in shape, size and function. It is the plan which is different, rather than the raw materials. In the same way, one might pull down a church and use the bricks and beams to build an old people's home.

But the astonishing fact that the genome can carry complete batteries of genes devoted to very different structural ends has not received the attention it warrants, perhaps because it was so difficult to fit into the primitive picture of genetics as it existed a generation ago. Now that we know there are regulator genes which control the expression of workhorse genes it becomes easier for the unimaginative to accept the idea.

From it arise new possibilities in interpreting the mechanism of evolution. There is an evident parallel between the transformation of tadpole to frog and the transformation of fish to amphibian. 'The frog tadpole is essentially a fish,' says Earl Frieden of Florida State University. Is it possible that the fish carries or carried a battery of genes specifying 'amphibian' in a suppressed state and that these were suddenly activated? And similarly for other major evolutionary advances. If so, a great many puzzling facts suddenly fall into place.

It becomes easy to understand why twelve mammalian lines began to exhibit similar characteristics. All were carrying the same, or similar, sets of masked genes which began to be activated about the same time, perhaps because they were triggered by the same environmental circumstances. All instances of parallel evolution become understandable.

We have seen that molelike creatures, almost indistinguishable anatomically, developed independently in Asia and in South America. Is it not easier to believe that they did so by the unmasking, in similar situations, of similar genes than to suppose that the same group of mutations occurred twice, in different places, by pure chance? And when we are asked to believe that it occurred four times, as with the anteaters, credulity fails.

Again, the concept enables us to understand why so few intermediate forms are found. If the transformation occurred very rapidly, not necessarily in a single generation but at least rapidly on the evolutionary scale, the chance of transitional forms encountering the geological conditions necessary for preservation would be enormously reduced.

When the mammals decided to return to a marine existence in the form of dolphins and whales, the fact that they developed flippers very like those of the ichthyosaurus, which existed in the Mesozoic long before, is explicable as a new unmasking of the genes which had in the interim been suppressed.

That genes should be preserved, unutilised, for millions of years instead of gradually vanishing or mutating to some quite different form may seem strange, but we have proof that this can happen, in the fact, already mentioned, that primitive organs, such as gills, reappear for a while during the course of the embryo's development.

Here, once more, we see the unmasking of genes long disused, though why this 'recapitulation' of earlier forms should occur remains wholly mysterious. It would seem that suppressor genes are subject to their own necessities.

'Masking theory', as I shall call this notion, also helps to explain the periods of evolutionary quiescence and bursts of variation ('radiation') which we discussed earlier. Obviously, if genes are suppressed, no amount of mutation or rearrangement is going to cause changes in the phenotype. On the other hand, if mutations have been stored up over a long period and are suddenly derepressed we should expect a flowering of different varieties.

Tucked away in the files of biological periodicals and long forgotten, are scores, perhaps hundreds, of observations which were regarded as wholly mysterious but which can be explained by masking theory. For instance, there is the spiny lobster, *Palinurus*, known to the French as *langouste*, whose eyes are located on the end of stalks. If you cut through the stalk a new eye regenerates. But if you cut at the base of the stalk, which removes the associated nerve-ganglion, what regenerates is not an eye but an antenna.

Then there is the puzzling frog discovered in the mountain meadows of New Guinea, which Professor Etienne Wolff described in 1971 under the rubric 'The Big Problems posed by a little Frog'. Unlike all other frogs, which lay eggs, *Nectophrymoides occidentalis* brings forth its young alive. That implies a womb, a placenta, a yolk sac and other modifications, for which it must possess the genes. Are they present in other frogs, unactivated? If so, why have they been activated in this one species? On the evolutionary scale, it has jumped a few million years. Similarly, there is just one viviparous earwig. Or again, only one shark in the genus *Mustelus* has a placenta, although they all live in an identical environment, namely coastal waters.

Such cases must mean one of two things, either fatal to the concept of a slow accumulation of variations. Either the same mutations occur repeatedly (in which case they can hardly be due to chance), or the genes are there all the time, but are unmasked in appropriate circumstances.

3 Infant Precocity

There is one more facet to the story and it is a curious one. You will recall that the fishes are supposed to have arisen from the group of organisms containing jellyfish, sea anemones and corals – creatures which have no blood, no excretory system and which remain rooted to one spot. They do, however, produce a free-swimming larval form with a primitive backbone, and other signs of incipient fishdom. This larva settles down after a few days on a piece of rock and converts itself into a jellyfish, anemone or coral. Thus we here see – in contrast to the frog – evolution turned backwards. At metamorphosis a more advanced form reverts to an earlier stage. Such functional return to an earlier state is in fact quite common.

Once again, we are forced to suppose the existence of two sets of genes, or two gene plans, the more advanced being briefly unmasked and then repressed again. The question of how the Cnidarians (as this group of organisms is known) accumulated a set of genes which they hardly use and which can scarcely be of evolutionary advantage to them, is even harder to answer than in my earlier examples. However it does suggest very strongly that new genes are not accumulated by simple processes of natural selection in the manner always assumed. It also confirms that such gene sets can remain latent, without being dissipated by chance mutations, for prolonged periods.

Sir Gavin de Beer explains the eventual emergence of these latent genes in the form of fishes by supposing that sexual reproduction, which normally occurs in the adult Cnidarian, is somehow brought back to the larval stage, after which the larva continues to propagate itself as a larva without ever proceeding to become a full-fledged Cnidarian. How this could occur, or why, is not even hinted at. It is one more piece of wild evolutionary guesswork. This supposed process is known as paedogenesis.

In sum, if we assume the essential validity of masking theory, we can clear up many of the most puzzling features of evolution, but at the price of being faced with several unanswered questions namely (1) how are latent genes, or gene-plans, accumulated? (2) what causes them to be unmasked at a given moment? (3) why are the latent genes not dissipated by random mutations when they are maintained over long periods of time?

Darwinian theory offers an answer, which is implausible and unsubstantiated, to the first question – namely, that the new genes arose by chance and were preserved because useful – but is unable to reconcile it with the many evolutionary facts which I have cited, or with the facts of metamorphosis.

Molecular biologists are hypnotised by the potentialities of DNA. The invention of methods of determining the structure of DNA and RNA species quite rapidly has given their work a further impetus, and they often speak as if DNA held the whole secret of development. But, as we know from embryology and the transplant experiments I mentioned, genes are only expressed when the cellular environment calls for them to be expressed. The part of the cell outside the nucleus is essential, also, to DNA. Nuclei transplanted into cells with an unsuitable cytoplasm give up the ghost. The study of these external influences has lagged, but will certainly cause substantial modifications of the picture when it goes ahead.

At the end of it all, we seem to be no nearer finding an answer to the demands the evolutionists put upon the geneticists. That is, to account for the appearance of integrated groups of modifications at the crucial moment without disturbing the existing body functions.

Because they thought of the genome as 'a bundle of unit characters' rather than as an integrated whole, the difficulties of regulation were overlooked. This needlessness was intensified by selecting trivial instances, such as beak shape, and ignoring the complexities of body chemistry or organised form. To bring about even such a simple change as green colour in an insect demands many genes. (Eye colour in *Drosophila* depends on thirteen genes.) How many more are needed for the thirty or more reactions which are involved in making blood! It is rather as if a composer should decide to rewrite the woodwind passage in a symphonic composition. He must not only orchestrate the entry of the various woodwind instruments in relation to one another; he must also pay heed to what the rest of the orchestra is doing at the same period in time. Today the problem is at least recognised. The neo-Darwinians did not even perceive its existence.

That these sequences of coordinated reactions – and there are literally thousands of them in the human body – should all have arisen by chance mutation of single genes is in the highest degree unlikely.

It is as if we expected the famous monkeys who inadvertently typed out the plays of Shakespeare, to produce the works of Dante, Racine, Confucius, Tom Wolfe, the *Bhagavad Gita* and the latest copy of *Punch* in rapid succession. Moreover, the curious bunching I have described under the term 'radiation' would suggest that they produced one of Shakespeare's plays, then waited a million years and produced the remaining nineteen in an almost continuous stream. As Professor Grassé has said, with a welcome change of metaphor, 'The probability of dust carried by the wind reproducing Dürer's *Melancolia* is less infinitesimal than the probability of copy errors in the DNA molecule leading to the formation of the eye.'

The story I have told so far has been confined to the last quarter of the history of life, the half billion years which followed the pre-Cambrian period and started with the appearance of the fishes. Obviously we must also look at the billion and a half years preceding this, misty though they are. They saw important steps in the evolutionary story, perhaps the biggest steps of all, notably the appearance of multi-celled creatures. On the face of it, there seems no reason why the world should not to this day be populated solely by amoebae and bacteria, algae and foraminifera, fungi and other protozoans. Does Darwinian theory throw any light on developments during the springtime of evolution? And does it throw any light on the origin of life itself? These are the questions I shall examine in the next two chapters.

CHAPTER TEN

The One and the Many

1 *Evolution's Biggest Step*

With the exception of one or two isolated finds, the fossil record virtually ceases about 870 million years ago. As we push back into the misty period beyond we come across a discontinuity, a major shift in the evolutionary story, which for me is the most baffling of all the Darwinian puzzles and the one which most clearly exposes the inadequacy of natural selection as the sole arbiter of evolution. Professor S. M. Stanley considers that 'it is widely regarded as the foremost unsolved problem of paleontology'. I am thinking of the appearance of multicellular creatures – Metazoa – after a long period in which the only life-forms had been unicellular.

Until this point, the sea had been a soup of microscopic forms, it is assumed, much like those we find in pond-water today. Then, one day, no one knows when, some of them were seized with the idea of coming together – at first temporarily, later permanently – and forming colonies. They did not assemble simply in clotted masses or chains, but formed neat and purposeful patterns. In addition, they began to specialise. Some discovered how to contract and formed muscle cells. Some learned how to transmit electric impulses and became nerve cells. Others became the precursors of skin and bone. Moreover, the overall structure of the cell mass became increasingly differentiated into organs of various kinds. It was an astonishing development.

Presumably about the same time the first multi-celled plants – Metaphyta – arose from unicellular algae and other forms capable of utilising the energy in sunlight to synthesise sugars from the water and carbon dioxide surrounding them.

It is difficult to see anything in the life of the Protists, the earliest unicellular creatures, which might have indicated to an intelligent observer that such a revolution was imminent. Once the change had taken place, it is easy to see that natural selection would favour it, and that the Metazoa would proliferate. Being larger, they could readily consume the Protists as food, and superior internal organisation enabled them to make better use of it – that is, to extract more energy from it. But why should the process ever have started?

Because of the absence of fossils we do not know what the earliest metazoans looked like, or into what class they fell; nor do we know from what groups they evolved, though many guesses have been made. In these conditions it is possible to mount long controversies of the kind in which some scientists delight. Since they can never be resolved they can agreeably fill out a lifetime. Or, as Professor Bonner says in a classic understatement: 'There is some dispute among biologists as to the origin of the metazoa.'

For example, it is possible to put together an increasingly elaborate series of multi-cell algae from the basic unicellular *Chlamydomonas* through *Pandorina* to *Volvox* which comprises hundreds of cells. Our old friend *Hydra* is another strong candidate. Judging from the plurality of modern forms of simple organism, the probability is that the Metazoa evolved on several different occasions by different routes.

According to some authorities, the Metazoa arose not by the coming together of amoebid cells, but from the masses of jelly containing many nuclei, known as syncytia, which we even find today. Many fungi are syncytial. *Opalina*, for instance, has several small nuclei, while *Physarum* exists as a separate amoeboid but sometimes fuses with many others to form a plasmid containing thousands of nuclei.

Not only do we not know anything definite about the origin of the Metazoa we do not even know the date of this crucial event. The first known metazoan fossils are found in strata dated to 700 million years BP, when the fossil record effectively starts. But they are quite advanced forms and the origin must be much earlier. The finding in 1947 at Ediacara in Australia of a lode of fossils – 1,400 specimens of thirty species – gave us the first firm data about pre-Cambrian life. The find included primarily jellyfish and soft corals, but there were also traces of segmented worms and several peculiar creatures not known to modern taxonomists. These are dated 680–700 million years BP, but the burrows of worms have been found as far back as 1,000 million years BP.

Edinburgh's E. N. Clarkson comments: 'The first appearance of metazoan fossils is very abrupt. They are already differentiated into various phyla and it is not easy to understand the evolutionary relationship of the various kinds on fossil evidence alone.'

From the fact that a considerable variety of modern forms of simple metazoan exist, it is inferred that the metazoan pattern emerged several times in different forms; at any rate, colonial existence does not seem a rarity. Whether differentiation also occurred several times seems to me more doubtful.

2 *Colonial Government*

There are many cases where cells, each identical with the rest, live in colonies, or form strings by budding, from the algae onward.

But the classic instance, given in every biology text, is the beautiful *Volvox*, a sphere of photosynthesising cells which rotates rapidly under the influence of its cilia. These sometimes reverse their beat, so that the sphere spins the other way, and in this coordinated activity of the cilia we already see indications of some overall organisation. *Volvox* is strictly speaking an alga and is equipped with chlorophyll: a plant rather than an animal. At the appropriate moment some of the cells produce daughters which form their own sphere swimming within the first. Sometimes the process is repeated, so that grand-daughter spheres swim within the daughters. Then the mother sphere breaks up, its cells die, and the daughter colonies graduate to independence.

So what induces them to congregate? How come they form a

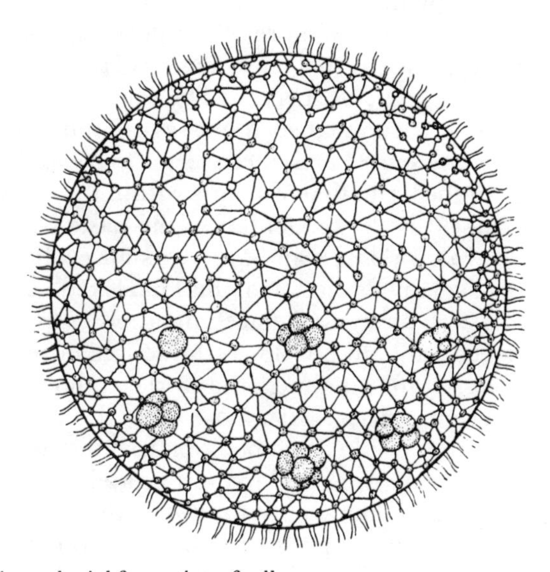

Volvox, showing colonial formation of cells

sphere and not just a loose bag or shapeless association? What mechanism makes the links? Darwin may tell us that the mechanism had selective advantage, but how in Heaven's name did it start?

There are even faint signs of specialisation, for *Volvox* has a distinguishable upper and lower pole. However, single cells, detached from the sphere, swim away quite happily, though they do not form colonies. This is rather odd since single cells from kindred species such as *Euglena* do form colonies. But then again the family nearest to the *Volvox* group, the Chlamydomonaceae, rarely forms colonies at all.

But the most intriguing gambit used by *Volvox* is to turn inside out. One side caves in, like a tennis ball on which you press your thumb, until it meets the far wall. A hole forms, through which the organism pours itself until it is completely everted. The suggestive thing about this manoeuvre is that it closely resembles what happens in the early development of the egg in almost all species up to man. The original cell divides to form a hollow ball. One side caves in ('invaginates') until it meets the opposite wall where it forms a hole. But then the process stops and we are left with a creature with an internal tube or gut joining mouth and anus. Meanwhile the mouth end has grown somewhat to narrow the opening into a mouth. Between the gut and the outer wall there is now a cavity, know as the coelom, in which in due course various body organs will form, notably those for sexual reproduction. Thus *Volvox* seems well on the way to forming the basic body shape common to a vast catalogue of animals (which is odd, of course, since it is a plant).

However, this seems much more plausible a step than is claimed by the rival theory; namely, that loose masses of protoplasm containing many nuclei are gradually pervaded by cell walls until they form a tissue. Though it is true that such masses of multinucleate protoplasm are formed by some fungi, there seems to be no sign of overall patterning nor of the formation of spaces within the mass.

Whether or not it is the progenitor of the basic body-pattern of animals or not, *Volvox* poses many questions for evolution. What is the advantage of colony formation of this kind? And how could such an elaborate pattern arise by the accumulation of fine differences? 'We have not been able to pin down the source of selection,' says Harvard's Professor John Bonner, the world authority on colonial organisms. It seems easier to suppose that the genome already carried genes to specify the basic body plan and that they became expressed rather prematurely and inappropriately in an alga.

Volvox, of course, we know only from its contemporary form. It probably did not exist in the pre-Cambrian and even if it did its frail structure would have left no fossil impression. But it poses the question of colonialism very clearly, I think, just because the sphere dissolves again.

Today we also find cells linked into little rosettes: four cells in the middle, like eggs in a nest, and a ring of half a dozen round them. Perhaps the earliest colonial forms were more like this. Another modern form resembles a stalk decorated with a number of bells, like the inflorescence of a flower. This, too, could have been an early pattern of colonialism.

We can gain a clue about colonialism from the sponges, one of the very earliest lines to arise in metazoan existence. We cannot call it unsuccessful – quite the contrary – since the sponges are still very much with us, yet it never led on to more advanced forms. They never developed the essential body cavity which becomes a gut and which seems to be a precondition of further advance.

John Bonner has made his main subject of study another, perhaps more challenging, colonial organism, the slime-mould *Dictyostelium*, discovered in 1935 by Kenneth Raper. The slime moulds have both plant and animal characteristics and are often grouped with the fungi. Normally this organism exists as a loose association of amoebae, which feed on bacteria and divide every three or four hours. But at certain times, as at a given signal, the amoebae stream towards a central point and form into a sausage-shaped slug which begins to crawl about, and orients accurately towards light and heat. There may be as many as 40,000 amoebae in the slug. After a while, it orients itself vertically and the uppermost cells construct a stalk, stiffened with cellulose, which they secrete. The rest of the cells stream towards the top of the stalk, where they form a small sphere. Each of these cells now forms a spore, soon to be released, after which the stalk collapses and its members die.

Bonner showed that the streaming together was brought about by a substance, emitted by one of the amoebae, which he named acrasin. Later he showed that the erection of the pillar was regulated not by opposition to gravity, but by a gas. 'The next question is: What good does this do the slime mould?' says Bonner and replies to his own question: 'There is no certain answer because we do not even know the evolutionary significance of the aggregation of the amoebae into cell masses in the first place.' Assuming it is advantageous, the gas may help the creature store its spores free of contact with damp soil. A gas, probably the same one, appears to control aggregation. Such experiments raise inviting questions about how far other kinds of cell communicate by means of gases but throw no light on the central problem of why they do so.

Bonner recalls that he once found himself explaining his work to two Russian university rectors, who were obviously totally uninterested, until he wrote on the blackboard the words 'social amoebae'. The Russians were electrified with curiosity and soon were beaming at the idea that even one-celled animals could be so sophisticated as to form collectives. The origin of social behaviour is a topic we shall discuss quite fully in Chapter Twelve. For the moment let us note only that it appears very low down in the scale of development indeed.

In the burgeoning embryo too, cells undergo 'morphogenetic movements' which assign them to various positions in the nascent organism. These must have originated at the start of metazoan life and the manoeuvres of *Dictyostelium* should in some way be linked with them.

In experiments in which one slug was stained with a harmless dye and the front end was then grafted to the back of an unstained slug, the stained cells moved rapidly up the slug, like a band of colour, until they had reached the front. They 'knew' it was their job to be front cells. Front cells placed at the front, and rear-end cells placed at the rear, stayed put. In the amoeboid society all cells are created equal but some are more equal than others. But, as Bonner points out, it is the leaders that perish and the laggard masses which survive.

The tendency of cells to sort themselves out has been studied by A. A. Moscona of Chicago University. He was inspired by a classic experiment performed near the turn of the century by Henry V. Williams of the University of North Carolina. Williams pressed sponges through bolting cloth until they were dissociated into single cells. Whereupon the cells came together again and formed new sponges on their own initiative. Not until the 'forties did anyone take this further, but then Johannes Holtfreter of the University of Rochester showed that vertebrate cells taken – admittedly from very young embryos – and dissociated would reassemble themselves. In 1952 Moscona proceeded to apply these ideas to chick and mouse tissue. He found that dissociated kidney cells would reform into recognisable kidney tubules, which would soon show secretory activity. Similarly liver cells would reassemble in structures resembling the intact organ and would accumulate glycogen. Heart cells coalesced into rhythmically contracting tissue. And when he mixed two types of cell – say, cartilage and kidney-forming – they soon sorted themselves out according to their identities.

The behaviour of *Dictyostelium* is by no means unique. There are quite a number of organisms which go through a life cycle in which a fruiting body is formed, after which the individual cells revert to an anonymous existence.

With these few clues, and some others as vague, we must remain content. The long and short of it is that the mechanism of metazoan existence is not understood and we have no firm clues as to how it emerged.

3 *It Makes a Difference*

The metazoan life-style involves more than colonial existence, however. It also depends on the specialisation of cells for different functions, and on the permutations of body form which multicellular construction makes possible. The metazoan is thus more than the sum of its parts and lifts the whole evolutionary process on to a new plane.

There are, to be sure, signs of incipient differentiation in the Protists. *Volvox* forms 'gonidia' to give rise to its daughters – but if you raise the level of nitrogen in the water, any of its constituent amoebae can do so. The Myxobacteria form a 'fruiting body' – a cell-mass tightly packed with spores which are eventually released and in due course become bacteria themselves. Biochemical specialisation occurred when the plants learned how to carry out the photosynthetic reaction by which chlorophyll traps the energy of sunlight.

But if we ask what led to the specialisation into muscle and nerve, bone and secretory cell, we can only speculate. One thing at least is clear. As soon as the cell mass was so large that some cells were cut off from the external world, specialisation had to occur: raw materials had to be transported to these interior cells and their needs signalled. How could such a thing come about? Why did not the surrounded cells simply die, as happens when bacteria multiply?

A clue as to what may have happened has been noted by Professor E. N. Willmer of Cambridge University. Willmer is the world authority in tissue culture; that is, on persuading little isolated groups of cells or even single cells to grow on a drop of liquid which can be placed under the microscope. Cells, generally speaking, don't much care for isolation and have to be fed a nutrient juice extracted from other living tissue to keep them happy.

When Willmer placed minced up fragments of chicken heart in the drop of liquid hanging beneath his microscope slide and watched, he saw that, after a while, undifferentiated cells began to creep out, and these were of four distinct types. Most numerous was a cell capable of forming bone and muscle or secreting collagen, the fibrous material widespread in the body. Second, there were cells which tended to join up edge to edge in a 'pavement' and which formed membranes such as skin. Thirdly there were sensitive cells which went on to become nerve cells. Finally, cells which wander about like amoebae. Later, they specialise further. Regardless of whether he started with tissue from mammals, amphibians or whatever animal group, he always got these four kinds. He makes the interesting suggestion that they may have originated, prior to the Metazoa, in the protozoan stage, by the incorporation of four different types of protozoan cell in the first metazoans.

What is certainly evident, is that by the time the metazoans were formed, cells were already very biochemically sophisticated. They already knew how to perform many functions, such as manufacturing and secreting highly specialised chemical substances with distinctive physiological effects.

But much more puzzling than cell specialisation is the question of how body shape as a whole is determined. And how are organs formed? Here again we are up against a blank wall. Some biologists have stressed the importance of a body cavity between the gut and the other wall – the coelom, which I have already mentioned. Filled with water it can give the body a certain rigidity helpful in burrowing, for instance. Later it becomes the site for internal organs, such as the heart and liver.

Another development which appears on the scene very early is segmentation: the repeating of a body unit, such as we see very clearly in millipedes. Most metazoans are divided into segments to a greater or lesser degree. We ourselves have repeated ribs and vertebrae, faint traces of our segmented ancestry. In a creature like the snake such repetition is very obvious. We can speculate that a gene-group, or master gene, responsible for a whole body segment, was in some way induced to perform its trick several times running, or even seventy or eighty times in a few cases.

It is tempting to refer the origins of such a pattern to budding processes such as we find in bacteria, algae and even rather more advanced forms such as *Microstomum*. Such reduplication would conduce, I imagine, to the eventual formation of a head and tail, in place of the radial symmetry found in jellyfish and starfish. But Liverpool University's R. C. Clark, referring both to the coelom and to segmentation, says sadly: 'There is no generally accepted theory of either.' None of the theories, he points out, demonstrates any selective advantage.

We have one last resort: to look at how the egg develops into an adult. Here, too, we find the formation of organs and the emergence of a body-plan. Essentially, ontogeny (as this kind of development is called) remains as mysterious as phylogeny (evolutionary development), but one fact of general significance has emerged. The instructions emanating from the nucleus are only carried out by permission of the cytoplasm. Here I shall cite Norman McLean of Southampton University: 'Selective gene activity does not itself explain differentiation.' And he adds, concerning the cytoplasmic factors controlling expression, 'Their identity remains a mystery.' (It is amusing to recall that the great geneticist, T. H. Morgan, announced in 1926: 'The cytoplasm can be ignored genetically.')

Finally, there is the problem presented by those creatures which produce larval forms which later metamorphose into the adult form. Here ignorance is even more profound.

4 Ancestors of Man?

Having made this tremendous evolutionary leap, how did life exploit its new possibilities? It shot off in every possible direction. By the late pre-Cambrian there was probably a greater variety of life forms than at any other period in earth's history. It was a golden age. The oceans were turgid with plankton on which the much larger multi-celled organisms could feed without fear of themselves being eaten. Only the land remained inhospitable.

Unfortunately we know all too little about it, since the fossil record is almost non-existent. Most of the time paleontologists have to infer from a few impressions in the mud, or the trail left by some unknown creature as it wriggled along the sea floor, or from a burrow preserved because it became plugged with a more durable material. When the general nature of the creature has been identified, we look at its modern representatives – if there are any – and infer what we can from them. But it seems that whole phyla existed which are now totally extinct. In those halcyon days, when almost anything which worked could make a living, it was not necessary to be efficient, for food was abundant; it was not necessary to be smart, for predators did not exist; which perhaps explains how the gastropods managed to develop and survive. It is an old biologist's joke that if Nature had not created Amphioxus, man would have had to invent it – for it contains such a rag-bag of key features that it can be fitted into anyone's theory to provide a link between two orders. But, as Alastair Graham of Reading University has observed, if gastropods had not existed, no one would have had the imagination to invent them. The gastropods have twisted the upper half of their bodies through 180 degrees and then flexed them through 180 degrees, so that the dorsal half lies back to front on the ventral half. It is as if a man should lay his shoulder blades flat on his stomach and stay that way for the rest of his life. No one can think of a convincing reason why such an agonising posture should offer any selective advantage, while it is rather easy to think of disadvantages, including the sanitation problem arising from having the mouth next to the anus. Yet gastropods form the second largest class in the animal kingdom. Some branches of the gastropod line later decided to unwind which didn't do them any harm.

Another very extraordinary thing about the gastropod is that it developed a penis. This was going to be extremely useful about a billion years later when it moved on to the land, but was strictly unnecessary in the sea. All other marine creatures manage very well by discharging eggs and sperm into the water at the same time. It is only on land that it becomes really necessary to make sure the sperm reach their destination. But perhaps such improbable ideas occur to one when one is standing on one's head and one's heels at the same moment.

Another group which got slightly ahead of the game was the cephalopods, the squids and octopi. They invented a powerful system of jet propulsion, enabling them to dart at their prey. This cannot have been much good to them in these early days when there was plenty of fodder and few organisms large enough to be worth darting at, but a billion years later it made them one of the most successful species in the Devonian ocean. Darting necessitated having good eyes and a brain to interpret what they saw. The cephalopods developed both: eyes on a model much nearer the mammalian eye than that of the insect, crab or trilobite, and a brain capable of making very delicate assessments of brightness but quite unable to comprehend weight.

Professor J. Z. Young of London University, who has spent most of his life considering the brain of the octopus, and his former student Martin Wells, who has done the same, say that the octopus is more like a cat or dog – a rather ill-tempered dog – than a primitive sea-creature. Professor Young has told me how, on entering the room containing the octopus tanks in the Naples Aquarium, an octopus had lain in wait and only when he was fully exposed at the tankside had drenched him with a great jet of water and then retired.

The octopus is also remarkably good at camouflaging itself. Its skin is spattered with little points which can be rapidly expanded to large dark areas, so that the creature can pass from pale grey through pink and red to black and back again. But however much it blushes or blackens its brow, its heart rate never alters. It is a cool customer. And it can also fool its enemies by ejecting a black cloud of ink in one direction while itself shooting off in the other. It can even roughen its normally smooth skin by raising thousands of papillae, if that will make it match its background better.

I've said enough to show that the cephalopods were in a class by themselves. With these advantages, they grew in size until some measured twenty-six feet across. How strange then that they never got further. At the same time, why did they get so far, at a time when the competition consisted of things like jellyfish and oysters?

Oysters, now. A muscle must have evolved to bend some flat and flabby proto-oyster into a purse, while simultaneously the chemical processes needed to secrete a calcareous shell and line it with mother-of-pearl were developed. It was not enough merely to secrete shell; it had to be formed in two neatly fitting halves, and a new muscle developed to force the two halves open when necessary. It doesn't seem very likely, but it happened and proved successful, as is shown by the fact that we still have oysters built on these lines today. But it never got further.

In contrast, the trilobites, which we have already examined, proved immensely successful, dominating the Cambrian seas numerically, only to die out entirely before the Mesozoic.

The $64,000 question, of course, was from which of these early groups did the vertebrates, and thus man himself, descend? There are as many opinions as there are authorities. Professor E. N. Willmer of Cambridge thinks they came from the Nemertine worms. Others say from the Annelida (e.g. earthworms), from the Mollusca, from the Arachnida or spiders, from the Echinodermata or starfishes and sea urchins, from the Hemichordates or the Urochordates (sea-squirts) or from the Cnidaria, which includes the corals and jellyfish.

Support for the Echinoderm theory is found in the fact that in some fossil species the exoskeleton is pierced by holes about at the point at which you might expect gills to form. Incidentally the echinoderms were perhaps the most variable of all these early groups. They assembled their armour plate in scores of fancy, imbricated patterns, like a fashion model trying on dress after dress. I mention this because it seems such a long way from the dour Darwinian picture of organisms precisely fitted to their niche in the struggle to survive. Many of these early creatures experimented with their entire body form, moving their mouth up to the end or down to the side, extending their feet or their neck, shifting their eyes (some had eyes in their tails, useful to tell you if you are truly out of sight in your burrow) and, as we have seen, twisting and contorting themselves. Far from being a struggle to survive, it looks more like one glorious romp.

As we saw in Chapter Three, the favoured answer to the question, whence the vertebrates? is from the larvae of proto-chordates.

5 *Sex and Sensitivity*

My survey has only skimmed the cream from the subject. Among the other questions seeking an answer is: how did cells develop irritability, that is, sensitivity to light, touch, chemicals and other stimuli? For, without this feature, they remain objects. Life can be defined by reactivity, and it is this reactivity which leads to brain, reason, consciousness and, perhaps, the sense of identity.

Another question is: how did sexual distinctions and sexual reproduction emerge? And how far did natural selection bring this about?

Sex is a theme which runs throughout the plant and animal kingdoms from the simplest bacterium to man, dictating bizarre courting behaviour and providing the ultimate motive behind food gathering, fighting and even play. We are as much a slave to its impulsions as is the amoeba and the distinctions between the sexes are the subject of passionate discussion even in this cool and libertarian age. How did this titanic force enter the evolutionary story and why?

It first appeared on the scene when a bacterium nestled alongside one of its peers and injected into the latter a portion of its own DNA. At this stage, you note, sex (meaning genetic recombination) was not associated with reproduction, nor were the sexes distinguished. The theory is that recombining the genetic material was potentially advantageous since it created the possibility of a more efficient genome emerging. Two useful mutations might be brought together. True, two harmful ones might be brought together too, but the success of the former would outweigh the failure of the latter. On this basis, the sex practised by bacteria was of minimal advantage. The donor gained nothing, and it was very much chance whether the recipient's additional bit of DNA was going to be helpful. Not until the eukaryotic cell arose, with its careful splitting of the genetic material into two halves, which then were swapped, did sex provide a reasonably effective device for recombination and not until then did natural selection exert any significant effect.

So we may wonder why the mechanism emerged, and, having emerged, how it was perpetuated. What impelled that bacterium to snuggle up to its neighbour? How did it acquire a mechanism for ejecting some (but not all) of its DNA? Was the recipient really more likely to survive? The whole thing is utterly improbable. Unless there was some mechanism compelling bacteria to act thus repeatedly the system could not have got going at all. Then, whence the mechanism?

But let us swallow the improbability of bacterial sex and ask whether, even in its more sophisticated, modern form, sex is really of evolutionary advantage? Many biologists doubt it. Professor G. C. Williams of Princeton has put his doubts into a book which he introduces with the words: 'This book is written from a conviction that the prevalence of sexual reproduction in higher plants and animals is inconsistent with current evolutionary theory.'

The first question which presents itself is: if sexual reproduction is so wonderful, why are there such a large number of asexual forms – organisms which either fertilise themselves, reproduce parthenogenetically (without a male) or manage by budding, spreading roots and so on? Most weeds manage without sex and are all too successful. In some organisms both sexual and asexual forms exist and in such cases the asexual form is usually the dominant one. There are also forms which, normally asexual, revert to sexuality from time to time. The common dandelion, you will have noticed, has kept its showy insect-attracting flower in case of need, although it reproduces asexually. Also telling is the fact that sexual and asexual populations evolve at the same rate.

Since an organism reproducing sexually loses half its genes it suffers a 50 per cent disadvantage. On the other hand, it gains in flexibility. The asexual organism preserves its abilities but does not throw up the new forms which may be more successful when conditions are changing. Thus asexual reproduction is efficient in the short run, sexual reproduction in the long run. (This could be why some organisms switch between the two, though how they do so, and still more how they know *when* to do so, is deepest mystery.)

This brings us to the real, brain-busting problem: how can selection favour a mechanism which will only show benefits in the future, if at all?

Are we sure, in any case, that a high level of adaptability is really the advantage that it is claimed to be? Some biologists have argued that evolution proceeds too fast, and that the problem is to maintain stability (a point I have touched on before).

In short, there is a fundamental contradiction between the claim that loss of the power of recombination increases the likelihood of a species becoming extinct and the claim that it improves the species' chance of survival when circumstances are changing. This contradiction remains largely unresolved.

Where sexual reproduction obtains, mechanisms of a surprising nature may act to preserve the optimum ratio between the sexes. Normally this is 1:1, but where the male has (for instance) a double set of chromosomes, the female will outnumber the male, showing that it is genetic equality, not just numerical equality, which is at stake. It is curious too that the proportion of males born after a war in which men have been killed seems to rise. Among some fish, as well as simpler creatures, sex changes take place to restore an uneven sex ratio. (Examples: in the cat, 107 males to 100 females. In man, 104, in the sheep 98.)

Among the oddites is the marine worm *Bonellia*, long thought to exist only in female form. Eventually, much-reduced males – as many as eighty – were found within the oviduct of the female. The larvae of *Bonellia* are sexually neutral. They attach to the proboscis of the mother, which contains a substance which induces masculinisation.

Since the unhindered course of development, even in mammals, leads to the formation of a female, we may ask how and where in the course of evolution was masculinity introduced. Answer comes there none.

Hung up as we are on the idea that every individual is either male or female and remains so, it might do us good to contemplate the successive hermaphroditism of the appropriately named *Crepidula fornicata*. These gastropod molluscs pile up, one above the other, on an empty shell. At the bottom is a female, fertilised by a female who becomes male. Each member of the stack plays male to the one below, female to the one above, changing sex for the purpose as needed.

Professor Williams concludes his book on sex and evolution on a note of despair which we can, justifiably I think, echo: 'I do not really understand the role of sex in either organic or biotic evolution.' The Darwinians could afford to think over that one.

CHAPTER ELEVEN

The Origins of Life

1 Life Emerges

Of all the mysteries of biology, unquestionably the most baffling is the question of how life arose on earth. Now that we have plodded backward down the evolutionary ladder for some three billion or more years, we come at last to this, the *fons et origo* of the whole business.

Scientists have espoused two main theories. Either living entities floated in from outer space (leaving the question of life's ultimate origin beyond our reach) or it arose spontaneously. Chance, which has brought about so many miracles, we are told, achieved the greatest feat of all: from 'a fortuitous concourse of atoms' it conjured self-perpetuating machines. Despite its apparent improbability, scientists mostly prefer the latter explanation. To the historian both theories reek with irony, for the medieval belief in the spontaneous origin of life was supposed to have been put to rest by Louis Pasteur in the nineteenth century, amid scenes of bitter controversy, while in the eighteenth century the great naturalist Georges Buffon elaborated a theory of the origin of life according to which the air was full of minute living particles, which, collected by living bodies, appeared as sperm. It attracted much mockery.

Pendent questions arise. If life can arise spontaneously, perhaps it did so more than once. Or perhaps it arrived more than once from the stars. Either way, the neatness of the evolutionary tree would be disrupted. Is the whole array of living things, extant and extinct, derived from one supreme progenitor, or is life 'polyphyletic', as the jargon has it? The fact that all living things employ basically the same chemistry, and in particular the fact that the genetic code is everywhere identical, are taken to imply that it arose only once. But if the chemistry and the code are necessary consequences of the structure of the materials involved, then even repeated events might possess the same features. (Nevertheless there are biologists who remain unconvinced. For example, J. R. Nursall of Edmonton University has declared that life originated simultaneously in many parts of the globe and spread into various environments. He also claims that the various phyla arose simultaneously and independently from these primitive ancestors.)

And when did this epoch-making event occur? This is a question I shall discuss in more detail later, but certainly it was very early: perhaps after one billion years of the earth's four-and-a-half-billion-year history, when conditions were, one might think, still highly inimical to it. The earth's crust was only just forming; the central land mass had not yet separated distinctly from the sea. Great upheavals and sinkings were still taking place. The temperature was high – according to some authorities it was as high as 50°C (over 120°F), which is about as hot as a cup of coffee. The atmosphere must have been darkened by the sulphurous fumes of volcanoes and even the sea was laden with toxic substances. But only beneath the surface of the sea could life conceivably manifest, for powerful ultra-violet rays were pouring through the thin atmosphere and would have destroyed the complex molecules which form the basis of living material.

Darwin himself thought it was 'unreasonable' to try to probe the origin of life, so that it is ironic, once again, that precisely here his theory of natural selection has proved productive: biochemists have extended it to cover the evolution of chemical substances, especially those required by the life process itself. A clear-cut instance is provided by haemoglobin, which arose from a substance known as myoglobin and then diversified into four variants. Thus by comparing the sequence of amino-acids in the haemoglobin from different animals we can tell when they must have diverged, and construct an evolutionary tree which can be compared with the conventional one based on fossils. When, in 1936, a Belgian chemist, Edouard Florkin, proposed the idea of molecular evolution it seemed revolutionary. Today we have a wealth of new evolutionary evidence derived from it.

The modern conception of how life originated was proposed by J. B. S. Haldane in 1928. He put forward the novel idea that the atmosphere of the primitive earth was based, not on oxygen, but on ammonia (that is, nitrogen and hydrogen). This is known as a reducing atmosphere, since it tends to remove oxygen atoms. The next step came in 1934 when the Russian academician Aleksandr Oparin published his classic book suggesting that lightning and ultra-violet light might have synthesised biologically interesting molecules in the primitive ocean. (In 1929 his colleague Vernadsky had protested that any research on such a subject was 'useless, illusory, harmful and dangerous'.)

However, it was not until 1954 that anyone attempted actual research. In that year Harold Urey, in the US, set his student, Stan Miller, to attempt the experimental production of biotic molecules by simulating the supposed primitive conditions in the laboratory. An earlier attempt to synthesise such molecules using a particle accelerator had been made by Berkeley's Melvin Calvin, but had produced nothing of much interest. So there was considerable amazement when Stan Miller's experiment yielded sizeable quantities of amino-acids and other organic molecules after only a few hours of electrical discharges.

Soon after, Calvin's group synthesised adenine, one of the components of DNA, from a similar mixture of ammonia, methane and water. These successes released a flood of experiments.

The primitive ocean, Haldane thought, must have been a thick, hot broth, half as dense as chicken soup. Even so, it appeared necessary to have some method of concentrating and preserving any biotic molecules that might form, and it was suggested that they might have been absorbed on clays: if so, it could not have emerged until clays, the ground-down remains of harder rocks, had formed. This, as we shall see, presented a difficulty.

Oparin had stressed the point that these raw materials would have to be encapsulated in some way and had pointed to droplets, known as co-acervates, which can be formed by suitable mixtures, as possible progenitors of the cell. More recently, Sidney Fox, at Miami, has produced something much nearer a natural structure, which he calls a microsphere; it assembles itself from a suitable mixture of chemical materials. Once encapsulated, the newly formed DNA and enzymes would have survived far better.

Today more than two hundred co-acervate systems have been constructed, and this does not seem to be a major obstacle.

The fundamental objection to all these theories is that they involve raising oneself by one's own bootstraps. You cannot make proteins without DNA, but you cannot make DNA without enzymes, which are proteins. It is a chicken and egg situation. That a suitable enzyme should have cropped up by chance, even in a long period, is implausible, considering the complexity of such molecules. And there cannot have been a long time: for reasons I shall discuss later we can infer that life emerged rather promptly the moment conditions were tolerable.

The spontaneous formation of DNA – or maybe RNA – is more

likely. In California, Leslie Orgel, formerly a chemist at Cambridge, is trying to get nucleotides to self-assemble themselves into DNA. They do so, though depressingly slowly. One new unit is added every quarter of an hour or so, as against a fraction of a second in the functioning cell.

But the fact remains that, as Miller surprisingly showed, the raw materials of life do spring into being and do, more leisurely, spontaneously form more complex substances. Is this, then, the secret of life: that it is a necessary consequence of the structure of matter? It is a point of such importance that we must investigate it in a separate section.

Whichever way you cut the cake, you have to wait for some incredibly lucky change. Of course, once you have got a system going, in which the product stimulates the starting material to work faster, you are out of the wood. At this point, natural selection favours your system which becomes steadily more plentiful.

That such an event should occur, however, is in the highest degree improbable. The late H. Quastler, a prominent biochemist, calculated the odds against it as 10^{-301}, that is, ten followed by 301 zeros to one, i.e. virtually impossible. Another biologist attempted a similar calculation for the whole universe, on the assumption that there were 10^{20} planets on which life might appear. He came up with the even more discouraging figure of 10^{-415}, rising to 10^{-600} if a longer DNA molecule was required. In short, the mechanism falls short of plausibility by hundreds of orders of magnitude. 'Perhaps some fallacy in the concept of natural selection will give us the way out,' as Frank Salisbury of Utah State University wistfully adds.

More complicated patterns of interaction are also possible. Thus you could have two interlocking systems, each of which supplies raw materials needed by the other; or more than two. But this just increases the improbability of such a configuration appearing by chance.

So we may suppose that life started as a replicating DNA molecule enclosed in a 'microsphere' or bag, and little else. This has been called a 'progenote' – something too simple to call a cell. In the course of time, it is supposed, the progenote acquired better biochemistry, as the appropriate enzymes happened to form, until we have something like the so-called blue-green algae of today. Though long classified as algae, these organisms are now recognised as a form of bacterium, and seem to have appeared some three billion years ago. In the rumen or second stomach of ruminants, such as cows, are found highly simplified bacteria which do not seem far removed from such primitive organisms.

Although DNA itself is stable, most bio-molecules are maintained dynamically: that is, they have to be continually rebuilt and repaired. For this, energy is required. Living organisms do not use energy as industry does; that is, by exploiting temperature differences. The cyclic chemical processes by which they exploit the energy in light are a 'phenomenon that is qualitatively new and characteristic for life'. It is, in fact, more characteristic of life than replication or evolution itself. 'Nowhere outside a living cell have dynamic states ever been observed,' notes E. Broda of Vienna University's Institute of Physical Chemistry.

In the absence of oxygen, it is now thought that the first bacteria functioned by splitting methane, or marsh gas, and a class of 'methanogens' or Archaebacteria, now vanished, has been postulated but fermentation proved an easier solution and served until the appearance of the chlorophyll molecule, which made it possible to trap the energy of sunlight directly.

Recently, I must add, chemists have begun to doubt the proposition that the primitive atmosphere was reducing, thus throwing the whole subject into confusion. Others dreamed up alternative methods by which ammonia might have been generated. Yet others think methane rather than ammonia was the main constituent. Obviously, the scientific view of the origin of life is highly speculative.

2 *Life and Light*

What is life? How can we define it? Many definitions have been proposed, but perhaps the best criterion of life is its capacity for self-repair. With few exceptions, living material is maintained by constant effort. It tends to break down and has to be repaired. We see this marvellous capacity for self-repair – so different from the mechanical devices which man constructs – all round us. When we cut ourselves, the wound is repaired, often so well we would never know that there had been one. When we lose blood, the system replaces what is lost. The same is true on the microscopic scale. Thrust a glass needle into a cell, tearing the membrane and disrupting the interior structures. The cell continues to function and in a few hours is as good as new.

The efficiency with which organisms restore the original plan, when damage has occurred, has so impressed many observers that they have concluded that some exterior, vitalistic force was needed to explain the phenomenon. Their astonishment was not wholly naive – we still are largely ignorant of how the restoration of form is achieved.

But one conclusion we can draw: such maintenance requires a

source of energy. It is a point which, until recently, biologists have tended to ignore. It was not enough, therefore, for chance to bring about the formation of a replicating molecule or molecular system. Right from the start, there must also have been an energy source. Clearly this poses a problem. Two major innovations had to occur simultaneously. As Professor S. Granick of the Rockefeller University has put it, the first organism must have been a primitive energy conversion unit.

Now, in the last analysis, the energy on which all life runs is that of light – sunlight. Sometimes it is extracted directly, sometimes indirectly. Those organisms incapable of photosynthesis – as the tapping of light energy is called – must live parasitically on photosynthesising plants. The earth is flooded with a vast spectrum of ether vibrations, mostly emanating from the sun, from mile-long radio waves down to gamma rays 10,000 billion times shorter. In this spectrum there is only a narrow 'window' of vibrations which are visible to man and which we call light – those lying between 380 and 1,100 millimicrons in length. Atmospheric absorption mops up almost all the remainder. Now, by happy chance (if Darwin is right) living things have evolved pigments which resonate to wavelengths precisely within that band, and this is true both as regards vision and as regards energy. And this circumstance, so Professor George Wald of Harvard maintains, must be true throughout the universe. 'There cannot be a planet on which photosynthesis or vision occurs in the far infra-red or far ultra-violet because these radiations are not appropriate to perform these functions,' he says.

One might add that for organisms living in the sea – as the first life-forms did – the 'window' is even narrower, since the water filters out most of the red end of the spectrum and, as you go deeper, finally all except the blue. Moreover, the shorter wavelengths – the ultra-violet rays – carry so much energy that they unzip the long-chain molecules of proteins and nucleic acids, thus making life impossible. Hence an atmosphere which filters out such wavelengths is necessary before life can emerge. So it is fortunate that our sun radiates most of its energy in this essential band, and that earth slowly developed an atmosphere which would filter out the ultra-violet. In short, as you can see, the conditions favourable to the emergence of life are quite stringent. In contradiction to those who maintain that there are countless inhabited planets, it may be the case that Earth is one of the very few planets where conditions are suitable.

There is no need for me to go into the incredibly complicated chemistry of photosynthesis, which involves whole sequences of chemical reactions and is still not completely understood, despite intensive efforts. There is, however, one rather intriguing point.

Photosynthesis depends on a substance known as chlorophyll. It exists in four slightly different forms, but all are based on a ring-like atomic structure formed by four chemical units known as pyrroles. In the middle of the ring is found an atom of magnesium. The four varieties of chlorophyll differ only in the make-up of the tail which is attached to the ring.

Curiously enough – and the significance of this is not clear – haemoglobin, which transports oxygen within the body from lungs to muscles, is also based on the pyrrole ring, though in this case it embraces an iron atom, not an atom of magnesium. Furthermore, several enzymes and Vitamin B_{12} also embody this peculiar chemical conformation, which is found in practically all living cells.

Until chemical evolution had produced this highly specialised structure, photosynthesis was not possible. Now, the chemistry involved in synthesising the pyrrole ring has very recently been elucidated by a team working in the Chemical Laboratory at Cambridge. It turns out that there are two distinct chemical systems at work, each with its own enzyme. One builds the components, the other forms them into a ring. So here again we have an improbable

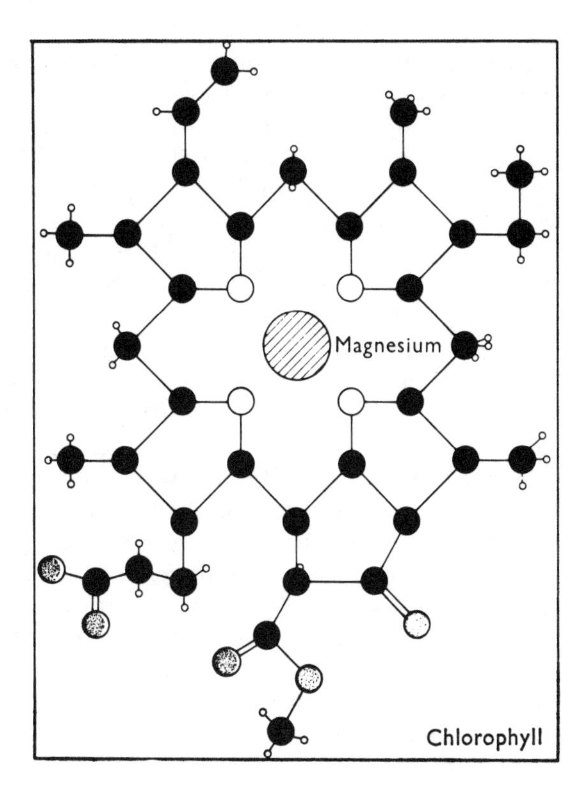

Diagram showing the structure of chlorophyll

coincidence. The formation by chance of one enzyme without the other would have been useless. Each complements the other.

The emergence of photosynthesis had the remarkable effect, over a long period of time, of changing the earth's atmosphere from one based on nitrogen to one based on oxygen. The former is called a reducing atmosphere, the latter an oxidising atmosphere. For creatures adapted to live in the former, oxygen is a poison.

In the reducing phase, most cells derived their energy from the simpler but comparatively inefficient process of fermentation; a few may have contrived to derive their energy from methane or marsh gas. The emergence of photosynthesis raised the efficiency of the organisms adopting it considerably and this may account for the sweeping success of the more complex nucleated cells. Further, as soon as the oxygen in the atmosphere rose above one per cent, the anaerobic organisms were handicapped; they were, not to put too fine a point on it, being poisoned. (It seems, then, that life as we know it created the conditions for its own existence, an occurrence more allied to purpose than to chance.)

Photosynthesis is a very sophisticated solution to the energy problem, and even fermentation is too complex to have arisen at the very start. How then did energy production begin?

Fundamentally, the problem is to split water into hydrogen and oxygen. Since iron has a mildly catalytic action, could the presence of iron compounds have played a role? Geologists are struck by the fact that early life forms are nearly always found in association with the so-called Red Beds – strata composed of limestone stained with iron. Professor Granick suggests that magnetite, an iron ore, may have been the original chemical unit which could decompose water, and that both chlorophyll and haemoglobin may have developed from inorganic precursors. This idea receives some support from the presence of limestone, which implies that the organisms were living in shallow waters. Burrowing in the sand could have been the way in which they escaped annihilation by the downpour of ultra-violet rays.

For the sake of scientific accuracy I must add that quite recently chemists have been changing their opinions about the nature of the primitive atmosphere. They are no longer convinced that it was formed of ammonia and methane, as Haldane believed, but rather that it comprised carbon dioxide, nitrogen and water vapour. One reason for this change of view is that the oldest known rocks, formed at least 3.8 billion years ago at Isua in Greenland, contain carbonates, which could only have formed in an atmosphere containing carbon dioxide. On the other hand, Professor Gerhard Schrauzer of the University of California at San Diego has shown that desert sands rich in titanium and iron can convert nitrogen and hydrogen to ammonia in the presence of sunlight. So the existence of an ammoniacal atmosphere cannot be ruled out.

It is, of course, the green of chlorophyll which gives its colour to the vegetation surrounding us today. In the same way, it lent its colour to the blue-green algae which were the original forms of life, after the bacteria. Indeed, they were bacteria, but bacteria which could photosynthesise. Later we find red, green and brown algae. These employ pigments other than chlorophyll to absorb light, pigments more suited to the wavelengths of light to which they are exposed, at the depth in the sea where they flourish. (If you have ever walked on a beach and picked up seaweed, you will have noticed these three colours: red, green and brown.) But these pigments, having picked up energy, transfer it to ordinary chlorophyll. Analogously, some primitive creatures have blood which is based not on iron-containing haemoglobin but on copper, so that they are truly blue-blooded. A few annelid worms boast green blood.

Where does all this leave us with respect to Darwinism? Even if we concede that natural selection was important in bringing about the dramatic change from living on ammonia to living on oxygen, it is significant that it could bring about such a complete change of mode. The doctrine that evolution proceeds only by gradual steps overlooks the fact that such steps can lead to a complete change of pattern.

Even so, it is very hard to swallow the idea that chance – or rather a long series of chances – built up such an extremely elaborate mechanism as photosynthesis, a mechanism which depends on substances far more complex than the raw materials which it transforms. Unless there was some inner necessity, some built-in, primordial disposition to consolidate into such a pattern, it is past belief that anything so intricate and idiosyncratic should appear.

The alternative theory, that life originated in space, does not avoid these difficulties. For, whereas in the nineteenth century, it was supposed simply that it was transported through space from the planets of some other solar system, in its modern form it assumes a spontaneous origin, on the same lines as that I have just described, but in interstellar space rather than on earth.

3 *Life from Space?*

Modern techniques of radio astronomy have revealed that many of the raw materials of life, such as ammonia and methane, are created in space by the action of ultra-violet light. On the strength of this, Professor Fred Hoyle, the cosmologist-mathematician-cum-science-fiction-writer, with his colleague N. Wickramasinghe, has argued that space, far from being the cold, lifeless place that astronomers used to describe, may in fact be pullulating with life.

The idea that life might arrive on earth from elsewhere is by no means new. It was mooted as long ago as 1865. In 1908 the great Swedish chemist, Svante Arrhenius, suggested that the radiation pressure of starlight could drive living spores through the universe like chaff before the wind. But as scientists came to understand the lethal characteristics of radiation, they realised that such spores would never survive on their own; embedded in meteorites, however, they might possibly do so. The British physicist Lord Kelvin firmly believed that life was embedded in rocks throughout the universe and was confirmed in this opinion by the finding of carbonaceous material in certain types of meteorite known as chondrites.

Only a few chondritic meteorites have been found, and these could easily have been contaminated by earthly life after their arrival. In 1959, new techniques were employed to examine meteorites and revealed hydrocarbons of the sort associated with organic substances such as butter. Then in 1961, Bartholomew Nagy, a Hungarian working at the University of Arizona, claimed to have found 'organised elements' resembling fossil algae in considerable quantities – too great to be explained by contamination – in two meteorites which had fallen in the nineteenth century in France. Alas, they turned out to be fakes made to embarrass Pasteur, a few weeks after his passionate defence of life as due to divine creation. But a search of other chondrites revealed lifelike forms, including a series of objects never seen on earth: five- and six-sided box-like structures containing minerals. Were these some form of life unknown to us?

Reviving the discussion in the 'seventies, Fred Hoyle proposed that life arrives in the form of bacteria embedded in cometary material. He thinks it is still arriving. Bacteria are so tough that they can withstand the extreme cold of space and also the destructive ultra-violet and even X-radiations. And it is certainly hard to see why earthbound bacteria should have evolved such resistance. There are in fact two main kinds of bacteria, cold-loving and heat-loving. The peculiar thing is that the cold-tolerant bacteria are often found in tropical soils! Why, Hoyle asks, should they have evolved the ability to multiply at temperatures below freezing, temperatures which are never met with in the tropics?

If it is true, as had been claimed, that life-forms existed 3.8 billion years ago, there simply was not enough time between the cooling of the earth and this date for them to have evolved. 'And even if the chemical building blocks were supplied . . . in some terrestrial pond,' insists Hoyle, 'Their assembly into life would be well-nigh impossible. The Earth is too small, the available time-scales too short, and life is far too complex for this to have happened.'

Hoyle has explored the chemistry and considers that there never was a 'primeval soup' and that the atmosphere was always oxygenated; he points to the lack of carbonaceous deposits in early rocks in support of this view.

But Hoyle's claim that the interstellar clouds – long a puzzle to astronomers – contain bacteria is pooh-poohed by astronomers. The absorption of radiation which Hoyle thinks is caused by bacteria is better explained as due to silicates, they say. Hoyle, however, calculates that the total mass of bacterial cells throughout the galaxy along with their associated viruses must amount to some ten million times the mass of the sun – an awesome thought!

He believes the arrival of bacteria, comet-borne, explains some disease epidemics and also cases where the same disease has broken out simultaneously in widely separated places; however, medical epidemiologists pooh-pooh this notion. Certainly the way in which some diseases, such as smallpox, remain quiescent for hundreds of years, only to flare up and decimate whole populations, is very hard to understand. Hoyle also cites the influenza epidemics, such as that of 1948. A Sardinian doctor, Margrassi, commenting on this epidemic, writes: 'We were able to verify the appearance of influenza in shepherds who were living for a long time alone, in solitary open country far from any inhabited centre; this occurred absolutely contemporaneously with the appearance of influenza in the nearest inhabited centres.' Sceptics, no doubt, would wish confirmation from observers better known than a Sardinian country doctor.

On the other hand, the accepted theory leaves much unexplained. For instance, molybdenum is a trace element essential for life, yet in nature it is so extremely rare (two parts in ten thousand) that it is hard to see how living things became dependent on it. Contrariwise, chromium is relatively abundant in nature, at 316 parts in 10,000; yet it is not utilised. Or again, why are there no traces of the primeval soup in the form of carbonaceous rocks of this epoch?

Francis Crick and Leslie Orgel, once at Cambridge but now at the Salk Institute in California, have considered the possibility that life is deliberately transmitted to chosen planets by advanced civilisations elsewhere in the universe. There seem to be no insuperable technological obstacles to so doing, they say, and add cynically that we ourselves should soon be able to pollute other planets in this way.

One objection to the idea that other civilisations are deliberately seeding our planet with life, or have done so in the past, is our failure to pick up any radio signals which would seem to have been transmitted by such other cultures. California's John Ball, however, accounts for this by the suggestion that our solar system has been set aside as a sort of zoo, in which advanced visitors can study such primitive forms of life as ourselves. Perhaps, however, the idea of quarantine, rather than of a zoological park, would be an even better explanation.

A billion years after the origin of life, by whatever means, a second major discontinuity occurred: the appearance of a much larger and more sophisticated type of cell than the simple bag of replicating molecules which I have so far described. And this refinement, too, involved rapid readjustments of a kind which seem beyond the scope of chance.

4 Life Diversifies

For a long time, biologists assumed that the first big innovation after the evolution of the cell was the division of life-forms into plants and animals. Since, however, such terms are inappropriate when discussing microscopic cells floating in the sea, we must speak of 'heterotrophs' which feed on organic materials and 'autotrophs' which are able to synthesise sugars and other nutrients from simple raw materials. You, the reader, are a heterotroph and only continue to survive because you ingest plants, which are autotrophs, or animals which themselves fed on autotrophs. Most biologists believe that the first organisms must have been heterotrophs, living on organic compounds which had been formed without the intervention of life. When the supply of these began to run low, some cells learned the trick of making their own nutrients; before long, other cells learned the trick of living cannibalistically on them.

This simple two-part scheme was soon found to be inadequate; the fungi could not be fitted in. Though they display some of the characteristics of plants, they live by thrusting feeding tubes into decaying organic material and so are autotrophs. Biologists therefore opted for a three-part classification: plants, animals and fungi.

However, it now turns out that this diversification of life was far from being the most important novelty in the evolution of one-celled forms. By the 1950s biologists realised that a remarkable increase in the size and efficiency of such cells, which took place at some unknown date, was a much more significant phenomenon. The primitive cell, such as we see in bacteria, has no nucleus: the genetic material is a simple ring of single-stranded DNA. It is known as a prokaryote, from the Greek, *karyon*, a nucleus. Then there suddenly appeared on the scene a much larger, more efficient cell, termed a eukaryote, which has a nucleus: that is, its genetic material is enclosed within a membrane. The nucleus also contains an organising body known as the nucleolus and other specialised organelles. Outside the nucleus, in the cytoplasm, are yet other organs, notably the mitochondria, boat-shaped structures packed with enzymes, which serve to extract energy from raw materials and – in the case of plants – plastids which synthesise nutrients with the energy derived from sunlight.

The eukaryotic cell is much larger and more efficient than the prokaryotic cell – ten times as large in some cases. It contains much more genetic material, and this material is not a simple loop but is organised into chromosomes. At the time of cell division, the chromosomes are split down the middle and drawn into the daughter cells by a structure called the centriole, a process known as mitosis. This ensures that the daughter cells are identical. Moreover, when sexual conjugation occurs, half the genetic material in each cell is thrown away, (a mechanism known as meiosis) and the remaining material is merged. This ensures that the fertilised egg, and the adult which growns from it, contains a mixture of characters from each parent. In short, heredity as we know it, and the operation of natural selection, only become possible with the emergence of the eukaryote.

The contrast between prokaryote and eukaryote has been called 'the greatest single evolutionary discontinuity to be found in the present-day living world'. How this advance was achieved remains one of the major mysteries of biology. It was not simply a discontinuity, but, 'one of the greatest single advances in the whole course of evolution,' says Professor W. S. Bullough of London University.

How, then, did this crucial development occur? At first, most biologists assumed that eukaryotes evolved from prokaryotes. The doubts arose and now some believe they arose independently and perhaps simultaneously. Yet others suggest that the prokaryote is a degenerate eukaryote – an eukaryote which has lost its novel features. However, the fact that both these kinds of cell embody DNA based on the same code, employ the same biochemical pathways in photosynthesis, and exhibit many other biochemical similarities, strongly suggests a common origin, despite the numerous differences in structure, of which I have mentioned only the most important.

Recently another rather startling theory of the origin of eukaryotes has begun to dominate the scene.

The view now held by most biologists is that the eukaryotic cell arose by gobbling up bacteria with useful properties and incorporating them, a process known as endosymbiosis. The idea that plastids – the photosynthetic units found in plant cells – might have such an origin was first proposed in the 'nineties but was not taken very seriously. The idea that mitochondria, the power-plants of both plant and animal cells, might likewise have been incorporated was made a few years later and likewise ignored. But in the 'sixties, Carl Sagan, biologist and author, revived the idea, which was enthusiastically developed by his wife, Lynn Margulis.

Endosymbiosis is, in point of fact, not uncommon in nature, especially where bacteria are concerned. We ourselves have bacteria in our gut which perform useful functions. There are more than 150 genera of invertebrates known to harbour bacteria. An amoeba can flow round an object and absorb it or open small channels in its surface by which even smaller particles can be taken in, afterwards resealing the outer membrane – a process known as pinocytosis – so that there is no fundamental difficulty in accepting the idea. Thus the development of pinocytosis may have been the crucial step which made the eukaryotic cell possible.

The argument that mitochondria are the descendents of incorporated bacteria is quite convincing. Mitochondria can grow and divide. By the mid-'sixties it had been shown that they possess double-stranded DNA and can manufacture most of the proteins they need – but not all, having presumably found they could pick some of them up from the host cell.

But Lynn Margulis took the idea further, proposing that at least two other features of the eukaryotic cell had been acquired by endosymbiotic means. One was the flagellum, the whip-like structure which propels some eukaryotic cells. (Structurally, it is different from and more complex than the cilia of simple cells.) Noting that the Protist *Myxotricha* is propelled by spirochaetes attached to its surface, she proposed that flagella are incorporated spirochaetes. Finally in 1970 she completed the picture by suggesting that the centriole – the device which separates the chromosomes at cell division – was also of endosymbiotic origin. The eukaryotic cell, it seems, is an ogre who has enslaved no fewer than four other organisms to work for it.

Of course, this daring hypothesis leaves us with numerous problems. How, for instance, did the cell manage before it acquired mitochondria? Why is some of the old circular DNA left over in the cytoplasm? Do the plastids of plants and the mitochondria of animals have a common origin? And so on.

However, we need not pursue these technical points. All we need do is register the fact that Darwinian theory scarcely explains such an astonishing development. To be sure, to postulate endosymbiosis is not an explanation; it is simply a description. It offers no explanation of how meiosis appeared. It does not account for the appearance of novel structures such as the nucleolus, the Golgi apparatus, or the microtubules which distinguish the eukaryotic cell. Above all, it does not explain how DNA came to be organised into chromosomes and enveloped in a nuclear membrane.

In short, far too many things seem to have been happening at once for chance to be an adequate explanation, and we are left with an enigma.

Finally, there is the question of what followed the bacteria and the blue-green algae. That these were the first life-forms is little disputed, but what they evolved into is a matter on which there is no agreement at all.

The viruses present a particularly puzzling problem, since they cannot reproduce themselves except by taking over the genetic mechanism of a cell. Are they, as some biologists think, genes which escaped from cells and set up existence on their own? If so, how did they get out and how did they acquire their protective coat, to say nothing of the special enzymes which enable them to intervene in the cell's genetic machinery?

Of the four classes of Protozoa, no one can agree which is the most primitive. The nineteenth century assigned this status to the amoeba; today the flagellates are considered the more likely candidates. Perhaps the question is unanswerable: some biologists hold that all four groups originated independently. Once again we find, in this crucial case, that there are no phylogenies.

5 *Life-and-Death Matters*

Now let me turn to the crucial question: just when did these momentous events occur? Was there, in fact, really enough time for chance to operate?

The fossil record is sparse, but recent discoveries have filled in some vital gaps. Many of them were made in the previously unexplored wilds of Africa and Australia. Fifty years ago, the known pre-Cambrian fossils could have been counted on the fingers. Today a computer is needed to record all the specimens which have been examined. Even so, we are confined to half a dozen snapshots in the long history of earth.

The first step into this previously uncharted territory was taken, as long ago as 1850 when Canadian naturalists found rhythmically laminated structures in the marbles of Grenville Province and thought they were of organic origin. Sir William Dawson named the structures *Eozoon canadense*, but when a panel of eminent geologists was asked to vote on the question 'Is *Eozoon canadense* of organic origin?' at the International Geological Congress of 1888, the vote was 'No' nine to four.

Early in the present century, one Charles Doolittle Walcott, a distinguished paleontologist, unaware of this debate, was astonished to come across pillarlike structures up to a foot across in lake beds in north-west America. Walcott thought they might be the remains of mats of ancient algae, but other biologists were sceptical. Walcott died in 1927 and it was not until 1954 that Stanley Tyler, of the University of Wisconsin, and Elso Barghoorn, of Harvard, investigating similar structures in Gunflint cherts near Lake Superior, showed that they did indeed contain microscopic plants and algae. Dawson and Walcott had been correct in their speculation. Since then, stromatolites have been identified in many parts of the world, culminating in the finding of modern stromatolites, very much like the ancient ones, on the Australian coast; they are of organic origin, which clinches the matter.

In Canada, stromatolites occur in rocks from 1,600 million to 2,500 million years old; they are widespread and abundant. While there is compelling evidence that life existed 2.7 billion years ago in Canada, and slightly less compelling evidence that it existed 2.9 billion years ago in Africa, it is possibly much more ancient.

The oldest known group of pre-Cambrian sedimentary rocks is found in Swaziland, and includes a sequence known as the Fig Tree cherts, which is well exposed in the Barberton district, a gold-mining area. These cherts were laid down a good 3.2 billion years ago. In 1965, Barghoorn collected specimens from these formations and, back in Harvard, where W. J. Schopf joined him, examined them under the electron microscope. The two scientists were considerably astonished to find in their specimens rods with double walls unlike anything seen before. They named the organism *Eobacterium isolatum*, the unique-dawn bacterium. Later, they found larger fossils of spherical shape under the light microscope, also novel, and named them *Archaeospheroides barbertonensis*. Says Barghoorn 'The existence of these two organisms, successful inhabitants of an aquatic environment more than three billion years ago, is evidence that the first evolutionary threshhold – the transition from chemical to organic evolution – had been safely crossed at some even earlier date.'

More recently still – in 1979 – the probable date of life's origin was pushed back even further. A German and a French biologist, H. D. Pflug and H. Jaeschke Boyer, studied the ancient rocks at Isua in Greenland dating from 3.8 billion years ago. Initially, their interest had been attracted by the fact that these rocks were carbonaceous, implying that the earth's atmosphere was already laden with carbon dioxide, not nitrogen. But, to their amazement, they found something much queerer: 'cell-like inclusions' of apparently biological origin, which they named *Isosphaera*. This organism, which looks very like a modern yeast-cell, reproduced by budding. However,

yeast is a eukaryote whereas these cells must have been far earlier than any eukaryote. Ventures Pflug and his French colleague: 'Isosphaera may represent a half-way line between a microsphere-like protobiont and subsequent evolution.' In plainer words, a link between organic and inorganic matter.

Confirming this discovery, to some degree, a few months later an Australian postgraduate, J. S. R. Dunlop, working on his doctoral thesis, came across stromatolites dating from 3,400–3,500 million years ago, in the North Pole area of Western Australia. No micro-fossils were found in them, when a joint American-Australian team investigated, but they inferred that living things must have been around at that early date. However, some workers still disbelieve this.

So there is problem number one. The earth was formed some 4.5 billion years ago. It did not cool to reasonable temperatures until perhaps 4.2 billion years ago. Yet life manifested after only 15 per cent of its entire existence, when conditions were, one would have imagined, highly unfavourable. This does not give long enough for the series of happy chances which seem to have been needed.

Set against this the fact that eukaryotes did not appear until a billion years ago: indeed, Barghoorn thinks even this figure too high and would set it as low as 725 million years ago. (Certainly the fossils in the Gunflint cherts of 800 million years ago are prokaryote.) In that case, it took three billion years to get from prokaryotes to eukaryotes, as against less than half a billion to form life itself. That doesn't seem to make sense.

It has been claimed that eukaryotes are found in the Bitter Springs cherts, also in Australia, dating from one billion years ago, but Barghoorn does not find the evidence convincing.

Recent frantic mapping, especially of the Pilbara region, has uncovered further evidence of life about this epoch and E. G. Nisbet and C. T. Pillinger of the Department of Earth Sciences at Cambridge, reviewing the situation in 1981, concluded: 'It may be that as early as 3,500 Myr ago there was an abundant simple but diverse biological community in the seas.'

Here is another puzzling fact. The nature of the Isua rocks suggests that photosynthesis had evolved incredibly early. Certainly, photosynthetic bacteria were in existence as early as 3.2 billion years ago. (Barghoorn, also, advances strong reasons for supposing that the Fig Tree chert organisms were photosynthetic.) Yet the change from an oxygen-poor to an oxygen-rich atmosphere does not seem to have occurred until 2 billion years ago. Not only does this imply very rapid chemical evolution, since two or three simpler chemical processes are thought to have led up to the evolution of photosynthesis, but it also poses the question: if photosynthesis was such a great advance, why did many anaerobic forms persist so long? This puzzle is complicated by recent theories that the early atmosphere may even have been oxygen-rich. As the Smithsonian's Kenneth Towe puts it: 'The study of pre-Cambrian Earth history presents an enigma. On one hand, it is widely conceded that the primitive earth was initially devoid of molecular oxygen and that life originated in such an environment. On the other hand, many pre-Cambrian rocks, including the oldest known sediments, contain primary oxidised iron minerals which indicate some source of free oxygen at the time of their precipitation and deposition. Where did this free oxygen come from?' For several reasons, Towe rejects the idea that it came from photosynthetic bacteria and proposes that it must have come from dissociation of water molecules in the atmosphere by ultra-violet light. Perhaps the strongest of his arguments is the fact that (as I have mentioned) there are two reaction systems involved in photosynthesis, one of which only gets going when oxygen is present. Atmospheric oxygen thus may have primed the pump for photosynthesis.

The emergence of eukaryotes about one billion years ago is evidenced by the findings at Bitter Springs, and those at Ediacara (also in Australia), which date from about 600 million years ago. Already by two billion years ago, as the Gunflint chert series shows, life had diversified. Not only spheroids but filaments resembling the modern Oscillatoria have been found, as well as star-shaped objects, christened *Eoastrion*, and an umbrella-shaped object christened *Kakabekia*, unlike anything known in the modern world. Thus it was a matter of intense astonishment when this very organism was found alive and growing near Harlech in Wales, soon after. But by one billion years ago, sexual reproduction was starting, and evolution as we know it was at last under way. Why did diversification occur so late – after 85 per cent of earth's history?

Of course, many question marks remain. What kind of animal made the tracks and burrows found at Ediacara? And what was the creature resembling the three-legged symbol for the Isle of Man? Then there are certain curious oval impressions, which some say were caused by medusoids (jellyfish) settling down, but cynics say were merely the traces of gas bubbles. Because many of these microfossils are impossible to assign to known algal groups, they have been given the provisional title of 'acritarchs'.

Only one point need be added. The invention of life which could reproduce necessitated the regularisation of death. Only if less efficient forms were eliminated could evolution proceed. When did cells create the time-clock which knocks them off after twenty or so divisions? No one knows.

CHAPTER TWELVE

Inherited Behaviour

1 No Place Like Home

The collared dove normally lives on rocky slopes in the Middle East but in recent years it has started to invade Europe, spreading as far as France where it has not confined itself to its usual habitat but has spread into Alpine valleys and lowlands alike, and is displacing the wood pigeon.

Facts like this make nonsense of the Darwinian conception of creatures surviving the battle for existence by their superior adaptation to the niche they fill. The collared dove, it is evident, is equally happy in many niches.

Probably the ultimate example of such ubiquity is the common dandelion, which flourishes all over the world in hot and cold, wet and dry, fertile and sterile conditions.

Whether species survive or not, it must follow, depends far more on their behaviour than on the availability of niches. Their skill and persistence in finding food, their ingenuity in avoiding predators, their success in finding a mate and rearing a family – these are the determinants of survival.

Unfortunately, 'the behavioural mechanisms by which individuals select their habitat are not well understood' as the *Oxford Companion to Animal Behaviour* remarks with truly British understatement. The Swiss Nobelist, Jean Piaget, known for his studies of the mental development of children, was so intrigued by the problem of habitat selection that he went outside his usual field and made some animal studies. Thus he observed that an East European snail, *Xerophila obvia*, which was first seen in the Valais, the southernmost canton of Switzerland, in 1911, has subsequently been observed all over the place, including sites at high altitudes. Why, he wondered, does the small mollusc, *Vitrina nivalis*, which has a thin, fragile shell, choose to live at 2,500 to 3,000 metres (10,000 feet) where conditions are severe. Not because it has been crowded out from lower altitudes, where in fact it is rare. His third subject was the pond-snail, *Limnaea stagnalis*, which normally inhabits ponds and marshes. But the almost indistinguishable variant, *Limnaea lacustris*, chooses beaches and rocks. Why?

Ernst Mayr has pointed out an even more puzzling case – two varieties of barnacle, one of which attaches itself to rocks thirty metres below the surface, while its cousin prefers to do so at water level, where it is battered by wave action. When they reproduce, the larvae from both species congregate at a depth midway between these extremes before metamorphosing and deciding where to settle, which suggests that at some earlier period this was the preferred depth.

There are three types of tree warbler, which distribute themselves in the lower, middle and upper parts of the forest, respectively. They do not seem to be competing for food and, even if they were, this territorial urge would seem to offer little selective advantage. Or, to take an even more bizarre instance, there are in the Galapagos Islands crabs which appear to be covered with black fur – but the fur is actually composed of millions of flies, three deep. These flies come out at night and settle on the crabs, departing again in the morning. E. Curio, of the Max Planck Institute at Seewiesen in Austria, who discovered this fact, provides no explanation.

To my mind, the most extraordinary sample of behaviour – in the context of habitat – is that of certain wasps which, as Darwin relates, spend most of their lives under water, using their wings as fins. This wasp, *Polynema nutans*, lays its eggs mostly in the eggs of dragonflies and most of its life cycle, including copulation, occurs submerged. 'It often enters the water and dives about not by the use of its legs but of its wings,' Darwin relates, 'and remains as long as four hours beneath the surface. Yet it exhibits no modifications in structure in accordance with its abnormal habits.'

This convincingly shows that behavioural change precedes structural change, and one can well see how favourable to such change the behaviour must be. More recently, Konrad Lorenz has made the same point, noting how birds which make nodding movements while courting eventually develop highly coloured feathers or crests which draw attention to these movements – not the reverse.

The urge to find new habitats might be explained away as simply a search for some plentiful food supplies in some cases. But no such easy explanation can be advanced when we consider the truly mysterious urges of migrants. Salmon and eels return, not merely to the same area, but to the same river after their periodic trips to the Sargasso Sea. Why? The salmon, in order to make such a trip (or is it the reverse?), has to modify its entire physiology to do so. What is the origin of this compelling necessity? Swifts and martins return not only to the same area, but even to the same house, to build their nests, as do storks. If, as is often said, migratory behaviour is motivated by the urge to avoid winter cold and summer heat, it would surely be sufficient to return to the same latitude.

This homing urge becomes positively risky in the case of the green turtle which normally lives in feeding grounds on the mainland, as recounted in Chapter Four.

A final example, less well known, is worth citing. Newts, taken from a deep canyon, over a 300-metre-high ridge and deposited 4.5 kilometres away, found their way back to their usual abode within a few days.

In a larger sense, we can see habitat selection occurring when the fish became amphibians, when walking lizards became burrowers, or when the ancestors of whales decided to re-enter the sea, an example the penguin now seems to be following. There are, today, lizards which normally walk on their stumpy legs, but which, when scared, drop on to their bellies and wriggle away more rapidly than they can walk, a mode of progression known as reptation. They are on the way to becoming a race of burrowers, as has often happened before. But why are they seized with this strange urge? And why do some animals prefer a nocturnal existence?

In curious contrast, many creatures occupy much narrower ranges than their physiological limitations make them capable of exploiting. An extreme example of this is those mosquitoes which breed only in the water which collects at the base of the leaves of bromeliads, with the further requirement that they live in light conditions. Experiment showed that they could not breed in dishes of water taken from bromeliads: apparently they need cues from the leaves. Even odder, one species of *Kerteszia* (for that is their name) enters bamboo shoots by small holes and breeds inside – having lost the desire for open water space exhibited by most anopheles.

I have stressed habitat selection as the main way in which an animal 'selects' the selective forces which will act on it; but there are of course many other kinds of behaviour which affect its survival, including its efficiency at food gathering, escaping predators and reproducing its kind. Among these, mate selection is obviously important. Females choose males, in many types of animal, for the elaboration of their display or the loudness of their call. How, one asks, have they come to the conclusion that brightly coloured feathers or a booming cry are desirable? And why, sometimes, do hens mate, not with the victor, but with the cock which has been *defeated* by another cock? That doesn't seem very Darwinian.

While biologists have generally neglected the role of behaviour in evolution they have been prompt to pay lip-service to the notion. 'We greatly under-estimate the role which behaviour plays in evolution,' says Sir Alister Hardy. Ernst Mayr goes further. 'A shift into a new niche or adaptive zone is, almost without exception, initiated by a change in behaviour. The other adaptations to the new niche, particularly the structural ones, are acquired secondarily.' Even G. G. Simpson agrees that behaviour is a determinant of evolution as well as a result. But having paid lip service, biologists mostly turn back to their studies of natural selection as if behaviour were unimportant. The honourable exceptions are the ethologists, but they confine themselves to analysing specific aspects of behaviour, such as courtship, without facing up to the larger issues.

Darwin recognised the importance of behaviour and often commented on puzzling instances, but he made no attempt to integrate it with his theory. Nowadays, some Darwinians still regard behaviour as a result of evolution, rather than as a cause of it. Or, to be more precise, the animal affects the environment which then affects the animal.

Plants, in a modest way, exhibit behaviour too, preferring one habitat to another, developing barriers against hybrid fertilisation, and so on. The fact that they cannot move about severely limits their powers of selecting habitats, of course, and this may be why plant evolution is so much less complicated than animal evolution.

Behaviour is to some degree instinctive, to some extent learned. That part which is instinctive must be inherited. Does it then obey the laws of evolution and how does it originate?

2 *The Mystery of Instinct*

Biologists dislike dividing behaviour into learned and instinctive, pointing out that, in any practical situation, both types of behaviour are expressed. Nevertheless, the fact remains that some behaviour is purely instinctive. The chick which, while still encased in the eggshell, pecks to get out cannot have learned to do so. Moreover, it does not peck at random, but aims for just the part of the egg where an air-space has been provided.

The great French naturalist Henri Fabre recounts an even more telling instance. A certain beetle lays its eggs within the wood of a tree. When they hatch, the grubs spend three years wandering about, tunnelling within the tree. When they finally emerge, they build a complicated nest – never having seen such a nest before. Evidently they must carry within them an hereditary message of quite a detailed kind specifying what they are to do.

Biologists assume freely that such inheritance of specific behaviour patterns is possible, and indeed that it regularly occurs. Thus Dobzhansky roundly asserts: 'All bodily structures and functions, without exception, are products of heredity realised in some sequence of environments. So are all forms of behaviour, without exception.' This simply isn't true and it is lamentable that a man of Dobzhansky's standing should dogmatically assert it. Some forms of behaviour are, certainly; we have no way of knowing that all are.

But the plain fact is that the genetic mechanism shows not the slightest sign of being able to convey specific behaviour patterns. What it does is manufacture proteins. By producing more of certain hormones it could affect behaviour in an overall way – making the animal more aggressive, more passive or perhaps even more maternal. But there is not the faintest indication that it can hand on a behavioural programme of a specific kind, such as the sequence of actions involved in nest building.

If in fact behaviour is heritable, what are the units of behaviour which are passed on – for presumably there *are* units? No one has suggested an answer.

If behaviour can be inherited, presumably it can evolve, since natural selection will eliminate adverse or unsuccessful behaviours and preserve those which are advantageous. Biologists assume that when two species show similar patterns of behaviour this is because they both derive from the same ancestor. Yet Piaget has noted that the courting behaviour of closely related species of *Nesomimus*, living on adjacent islands in the Galapagos, varies from one island to another. If behaviour patterns are inherited, it must be a very weak process. On the other hand, a species of plover which formerly nested on beaches but now nests inland still paves its nest with stones, as it once used to when stones were all that was available – which suggests that behaviour can be very persistent.

The well-known case of the bower birds looks very much like the kind of 'overshoot' which we noted in the evolution of morphological features. These birds build towers as much as nine feet high, with internal chambers, in the middle of circular lawns. These they embellish with flowers, which they replace as soon as they are withered. They may dye the walls with the juice of berries, or decorate the bower with snail shells, bits of glass or spider silk. One bower contained, it was found, nearly a thousand pebbles, more than a thousand sticks and more than a thousand strands of grass.

G. Evelyn Hutchinson of Yale University, considers that the

birds' engineering, architectural and decorative skills, like their courtship displays, constitutes behaviour 'that in its complexity and refinement is unique in the non-human part of the animal kingdom'.

E. T. Gillard, Curator of the Department of Birds at the American Museum of Natural History, who has made a special study of these remarkable creatures, notes that some species of these birds have crests, but that the smaller the crest, the more elaborate the bower. The nests, he says, are simply 'externalised bundles of secondary sexual characteristics'. Interestingly enough, birds which do not have crests make the same head movements as those which do when showing off their bower to the female.

Often quoted as an instance of the evolution of behaviour is the case of a fly known as *Hilaria*, which manufactures a balloon out of silk and offers it to the female of its choice. The female accepts the balloon she likes best and then permits copulation. The origin of this curious ritual is claimed to be as follows. There are flies which present grubs, which they have paralysed, to the female. There are others which wrap the grub in silk before presenting it. But *Hilaria* discovered that it was not necessary to offer a diamond ring provided you offered the box!

This imputes to *Hilaria* a considerable degree of intellectual perspicacity.

When we ask ourselves how any instinctive pattern of behaviour arose in the first place *and became hereditarily fixed* we are given no answer and it is tempting to turn to a Lamarckian explanation. In the life of the individual, a successful response is at first consciously repeated but soon becomes habitual. How natural it seems that a line of animals should conserve what it has learned from experience. (Lamarck saw behaviour as determined exclusively by the environment, you recall. We need not follow him in this.) It is, after all, the way an animal behaves which makes a structural modification helpful or a handicap.

Ernst Mayr says: 'A shift into a new niche or adaptive zone is, almost without exception, initiated by a change in behaviour. The other adaptations to a new niche, particularly the structural ones, are acquired secondarily.' Most of the shifts which ethologists have actually been able to study have taken place without any structural modification.

In contradiction to this, we must note that language does not become innate, though this kind of behaviour has been repeated for thousands of years. (The contrary used to be maintained. James VI placed two children on an island in the Firth of Forth, with a dumb nurse, and in due course was relieved to find that they 'spak guid Hebrew'.)

When mutation was discovered, the role of behaviour as a selective factor was dropped from favour. It was only when, in the 'thirties, the new school of ethologists was born, that it was reinstated.

3 *Do As You Would Be Done By*

There is one class of behaviour which has long represented a major stumbling block to the theory of evolution, namely 'the vexatious problem of altruism', as Gould has called it. What is the evolutionary advantage of helping another to survive, especially if it involves risk to oneself? Recently, some biologists claim to have resolved this question. But have they?

Darwin said in 1859: 'Natural selection cannot possibly produce any modification in a species exclusively for the good of another species, though throughout nature one species incessantly takes advantage of and profits by the structures of others . . . If it could be proved that any part of the structure of any one species had been formed for the exclusive good of another species, it would annihilate my theory, for such could not have been produced through natural selection.'

Darwin was referring to structures, but his statement is equally applicable to behaviour, supposing behaviour to be inherited. Consequently, the occurrence of altruistic or unselfish behaviour, especially when it involves some risk or cost to the altruist, has long presented a challenge to Darwinism.

Certain small birds, such as robins, thrushes and titmice, for example, crouch low when a hawk approaches and emit a thin reedy whistle which warns other birds of the danger. From the viewpoint of the individual bird, it would presumably be wiser to remain silent and not give away one's position. A more dramatic example is provided by those bees which launch suicidal attacks on invaders of their hive. Darwin himself was puzzled by the case of the sterile worker bee, who accepts the extinction of his line in the interests of the queen; but a more general example is maternal behaviour – especially such features as the mother regurgitating food, half digested, for the benefit of the young.

Most generous of all, perhaps, are those ants which use some of their own eggs – so-called trophic eggs – to feed their queen and even other workers. We also find reciprocal altruism between different species, as in the case of cleaner-fish, or the cattle egret and the tick-bird, which warns the rhino on which it perches of the approach of foes. (But reciprocal altruism entails problems to which I shall come back in a moment.)

As Stephen Gould of Harvard admits: 'If the genetic components of human nature did not originate by natural selection, fundamental evolutionary theory is in trouble . . .'

In 1964 W. D. Hamilton, an English geneticist, published a classic paper which tackled this problem squarely and proposed a solution. Hamilton took the case of the social insects, and showed that the sterility of workers could be understood from the fact that such creatures have an unusual genetic mechanism which results in a worker's sisters acquiring more of its genes than its own offspring would – hence, it pays, in an evolutionary sense, to preserve the sister rather than the offspring.

Actually, J. B. S. Haldane had encapsulated this idea almost thirty years before during a conversation in a pub. When asked whether, as a believer in evolution, he would give up his life for his brother, he replied: 'For one brother? No. But for three brothers, yes . . . or nine cousins!'

Hamilton's paper did not explain the preservation of altruistic tendencies in creatures which do not possess the unusual genetic mechanism referred to. Nevertheless it emerged as a bulwark of theories about behavioral evolution. In 1975, Edward O. Wilson, Curator of Entomology at the Museum of Comparative Zoology at Harvard, made a powerful synthesis of the whole subject, dubbing it 'sociobiology'. The topic has been the focus of furious controversy, mainly because the central assumption – namely, that behaviour, and specifically human behaviour is, at least in part, genetically determined – is unacceptable to Marxists. With that aspect I am not concerned. The question which *is* relevant is whether such genetic explanations are factually correct, for, as Professor T. H. Frazzetta has said: 'If altruism arguments are correct, very many of our concepts about natural selection are wrong.'

I must therefore point to three major weaknesses or limitations to the otherwise impressively neat arguments of the sociobiologists. First, they define altruism quite arbitrarily as actions which increase the likelihood of survival for the recipient while decreasing them for the altruist. But the bird which regurgitates food, or the dolphin which supports an injured member of the group does not significantly reduce its own chances of survival. More importantly, the sociobiologists assume that there is a gene for altruism, or possibly an array of genes for particular altruistic actions (they do not make it very clear which). But this is sheer guesswork. There is, to come to the nub of it, a difference between altruism as the word is conventionally used and the purely genetic altruism of the sociobiologists.

In human societies, altruism may result from social conditioning, or from a thought-out position, or even from social pressures and conventions.

Wilson himself is confident that the existence of specific altruistic genes will soon be established. 'I also believe that it will be within our power to identify many of the genes which influence behaviour,' he declares. In contrast, Mayr maintains that 'it is no longer possible to assign an absolute selective value to a gene . . .' because genes operate differently in different environments. 'The genes are not units of evolution,' he maintains.

Be this as it may, the existence of altruism between different species – which is not uncommon – remains an obstinate enigma. A particular vivid example of this is the activity of cleaner-fish, which was studied by the late Conrad Limbaugh, of the Scripps Institute of Oceanography, until his sad death in a diving accident in 1960. While skin diving off the coast of Lower California in 1949 he observed the curious behaviour of golden kelp perch. This three-inch-long fish kept pecking at the sides of a wall-eye surf perch which had separated from its fellows. Intrigued, he began to collect other instances. In the Bahamas he found Pedersen's shrimp, a transparent creature, striped with white and spotted with violet. When a fish approaches it, as it hovers among some sea anemones, it waves its long antennae. The fish, if interested, presents its head and gill covers for cleaning. 'The shrimp climbs aboard and walks rapidly over the fish, checking irregularities tugging at parasites with its claws and cleaning injured areas. The fish remains almost motionless during this inspection and allows the shrimp to make minor incisions in order to get at subcutaneous parasites. As the shrimp approaches the gill covers, the fish opens each one in turn and allows the shrimp to enter and forage among the gills. The shrimp is even allowed to enter and leave the fish's mouth cavity. Local fishes quickly learn the location of these shrimp. They line up or crowd around for their turn and often wait to be cleaned when the shrimp has retired into the hole beside the anemone.'

Apart from the remarkable nature of the fish's behaviour, one may wonder what the shrimp gets out of these helpful activities.

Such altruistic behaviour had been noted as far back as 1892 (it was suggested that a spider which hitch-hikes on scorpions was actually removing parasites). Limbaugh's own work concentrated on a wrasse, known locally as *La Senorita*, which often has hundreds of 'clients'. He found that fish come long distances to be cleaned and attend at 'cleaning stations' where they queue for a place. Such behaviour, he commented, raises a great many questions for students of animal behaviour: for instance, why do ordinarily voracious fish refrain from devouring their little helpmeets?

The attempt to extend evolutionary theory to cover behaviour is a bold one – but, as it seems to many commentators, over-bold. Behaviour results from a complex mixture of learned and inherited factors, and the inherited factors may well prove to be subtler than people like Professor Wilson suppose. Does a bird sing solely in order to please its mate? Is it not possible (as Professor Donald Griffin of Rockefeller University believes) that it also sings for pleasure?

I should like to offer a final comment. If altruistic behaviour is truly inherited, man is in trouble, for he has probably ceased to transmit altruistic genes. The advantages of altruistic action, in the modern technological world, rarely accrue to his kin. They may very well benefit people who lack such genes. As the American baseball player Leo Durocher cynically remarked 'Good guys finish last.'

4 Done On Purpose

Much animal behaviour is quite automatic and produces stupid results if things go awry. For instance, where male and female birds take turns in sitting on their clutch of eggs, while the partner hunts for food, things go wrong if one bird is killed. The other, at the appropriate time, goes off in search of food and the eggs grow cool and die. It then comes back and sits on the inviable eggs for the due period, before going off again.

Similarly the wasp *Ammophilus*, which G. P. Baerends has described, places its larva in a hole and goes off to collect grubs on which it will feed when it emerges. But if each grub is removed while it is away searching for more, it will continue indefinitely to try to fill the cavity. Each phase of this activity is triggered by the one before and if it is not completed, it cannot get on with the programme.

Even stupider is the bird which cheerfully drops worms through a 'gape' or cardboard opening resembling a mouth. The fact that the imitation mouth is not backed up by a living offspring does not worry it.

Aside from such automatisms, we find behaviour which seems purposeless, though accurately planned. There is a spider which joins streams of ants. It has, of course, eight legs against the six which ants possess, but it gets over this by holding up its front two legs like antennae. The problem here is that there do not seem to be any predators from which it could be escaping. It is just climbing on the band-wagon.

Such phenomena make it all the harder to understand the astonishingly purposive and well-organised behaviour displayed in other cases. There is, for example, a crab which is equipped with a statolith – a small stone resting on sensitive hairs, the position of which indicates the crab's attitude or orientation to the vertical. If this stone falls out, it replaces it with another, or, in some cases, secretes a lump of stony material to replace it. It is hard to see how, when this situation first arose, the crab could have happened upon the correct way of restoring its balance organ.

Or again, consider the bola spider which, instead of weaving a web, hurls a thread terminating in a glob of material at its prey, for all the world like the bolas which Mexican horsemen swing round their head and launch at the legs of cattle they wish to capture. It seems inconceivable that, by pure chance, the creature both manufactured the weapon and discovered how to use it, for if its aim were defective at first, there would be no reward to condition it to repeat such behaviour.

Web-weaving spiders, observed by de Witt, neither run out of material nor end up with a surplus. Before they start to weave they contrive to manufacture precisely the amount they will need.

And how are we to account for the behaviour of the crab (known as *Birgus latro*, the thief) which leaves the sea, climbs a palm tree and collects a coconut? How, one asks, does it even know that there are coconuts to be had, much less that they contain delicious milk?

In many cases, complex behaviour is executed by organisms so small as to have very little in the way of brain. The larva of the caddis fly builds a case for itself, and if this is destroyed replaces it. If given too large a case by the experimenter, it adjusts it to the right size. That is to say, its behaviour is not completely stereotyped. Even sponges build complex geodetic structures of spicules to serve as a skeleton. Spicules of different lengths are selected and cemented together as if to a plan. And I have already recounted how *Microstomum* equips itself with stinging nematocysts, with endoderm, parenchyma and epidermis cooperating to manipulate them into the required position.

Such processes remain inexplicable and have driven many biologists to supposing that there must, somewhere, be a plan to which they adhere. As Sir Alister Hardy insists, they can hardly be the result of physico-chemical mechanisms alone. What is especially challenging is their anticipatory nature. How does *Microstomum* know that nematocysts will come in handy? Cicadas lay their eggs on twigs. The larvae which hatch from these at once drop to the ground and hide. How do they know it is prudent to do so?

'No mechanistic explanation can account for the orderliness and directiveness of organic activities,' remarks E. S. Russell in his famous book *The Directiveness of Organic Activities* – but this does not mean that there is an external directive force. Rather is it the case that living organisms have self-directing properties; but here I begin to anticipate the next chapter.

I incline to believe that there are still processes of communication underlying behaviour of which we have at present no more than an inkling. Let me quote an observation of Professor James S. Coleman of Johns Hopkins University, whose subject is social relations, since he puts it so vividly.

'Once when I was sitting on the edge of a cliff, a bundle of gnats hovered in front of me, and offered a strange sight. Each gnat was flying at high speed yet the bundle was motionless. Each gnat sped in an ellipse, spanning the diameter of the bundle, and by his frenetic flight maintaining the bundle motionless. Suddenly, the bundle itself darted – and then hovered again. It expanded and its boundaries became diffuse; then it contracted into a hard tight knot and darted again – all the while composed of nothing other than gnats flying their endless ellipses. It finally moved off and disappeared . . .

'Such a phenomenon offers enormous intellectual problems: how is each gnat's flight guided, when its direction bears almost no relation to the direction of the bundle? How does he maintain the path of his endless ellipse? And how does he come to change it when the bundle moves? What is the structure and what are the signals by which control is transmitted?'

I have often observed almost identical behaviour in flocks of birds. But birds, despite the opprobrious term 'bird-brain', have quite efficient brains, weighing several grams. Gnats, on the other hand, have microscopic brains comprising only a few hundred neurones. Professor Coleman offers no answers to the question he poses; neither do I.

CHAPTER THIRTEEN

Chance or Purpose?

1 Neo-Darwinism Assessed

Now that we have examined three billion years of evolutionary history, so far as speculation will take us, let us take stock. How does the official account of evolution stand up? In particular, how effective has the Darwinian theory of natural selection been in explaining what has occurred?

The short answer is that natural selection accounts very well for many of the minor changes in structure and this is true even at the biochemical level. This has been demonstrated in thousands of cases which have been minutely studied. Nevertheless, there are a surprising number of instances where the process does not seem to work. The most obvious is what I have called 'overshoot', as in the case of the peacock's tail, which makes flying difficult, rather a serious disadvantage for a bird. Why should peacocks be the one exception?

Less obvious but just as common is undershoot or even the failure to adapt at all. Darwin was particularly disturbed by such maladaptations as the bee which, by stinging, ensures its own death – a circumstance which, one might think, would secure elimination of the genes concerned with stinging.

This brings us to the second aspect of Darwinian theory, and the more important one, namely the origin of variations. Darwinians, as we know, assign variation to chance. Until recently, mutation was generally held to be the chief cause of such variations; today the consensus is that they are due to rearrangements, and now that we know about 'jumping genes' we see that drastic rearrangements are quite possible. Even so, this aspect of evolutionary theory is deeply unsatisfactory.

First, as the mathematicians have insisted, there is just not enough time for chance to evoke all the needed modifications, if single mutations are presumed to be involved and especially when several distinct mutations are needed simultaneously. Even G. G. Simpson, stout Darwinist that he is, agrees that to secure five mutations simultaneously would take forever. The idea that genes constantly jump about could evade this objection but leaves us with the problem of why they do so.

Then what is the explanation of the odd changes of tempo in evolution? In the last few million years man has diverged dramatically from the original stock, while the other primates have changed much less. Other species, such as sharks and frogs, have hardly changed at all. In some lines, evolution seems to have stopped. Darwinians have nothing helpful to say about this.

I think it is now clear that the Darwinian assumption that evolution was always gradual has failed, though this is not fatal to the theory.

But perhaps the most serious weakness of Darwinism is the failure of paleontologists to find convincing phylogenies or sequences of organisms demonstrating major evolutionary change. Naturally many will have escaped fossilisation or have been subsequently destroyed, but surely one or two should survive? The horse is often cited as the only fully worked-out example. But the fact is that the line from *Eohippus* to *Equus* is very erratic. It is alleged to show a continual increase in size, but the truth is that some of the variants were smaller than *Eohippus*, not larger. Specimens from different sources can be brought together in a convincing-looking sequence, but there is no evidence that they were actually ranged in this order in time. 'Phylogeny is still biology's major unfinished task,' admits Professor Hanson.

Moreover, as I have told, many of the mammalian sequences appear to have started up simultaneously. Parallel evolution is also common in the world of insects, notably with regard to the tracheae or tubes which convey oxygen to different parts of their bodies. 'There is no denying the parallel evolution of tracheae serving aerial respiration,' says Sidnie Manton of the British Museum (Natural History) in the mandarin language of her profession, 'which has taken place at least four or five times.' In Chapter Eight I offered an explanation of such parallelisms.

There are evidently things about the functioning of the genome which we do not yet understand. What is the range of possibilities available to it? Its repertoire must, in any given species, be limited. The logical deductions from the observable facts of evolution seem to me to be: (1) the genome must run through many permutations, discarding all those which are not inherently viable, and trying out the viable ones for effectiveness in the real world; (2) it must have a facility for storing such potential forms and for realising or activating them when circumstances demand a change. But in doing so it will eventually exhaust all the possibilities open to it, at which point its evolution will cease. The organism concerned will continue to flourish as long as conditions remain unchanged, but will become extinct if they change in a way to which the repertoire of stored variations cannot effectively respond.

These are bold conclusions. Some biologists still deny the possibility of storage, although the discovery of redundant DNA has made the first of the above assertions plausible. The postulated mechanism for release of variations is not known. However, a system of this kind would explain the changes in tempo, the sudden jumps which Eldredge has documented, the stasis of sharks and lungfish, the rapid evolution of man, and the appearance of atavisms and anomalous structures with much else. The genome must have its own logic.

Even then, the neo-Darwinian dogma leaves much unexplained. In particular, it is difficult to see why, when more adapted forms evolve, the less adapted ones frequently persist with complete success. If the Darwinian theory of competition were correct, they would perish. An excellent example of this, which gives even Darwinians pause, is the coexistence, for 100 million years, of fish with an evolved cheekbone and fish with an archaic cheekbone. As Professor J. P. Lehman observes: 'It therefore looks as if the arrangement of the cheekbone was irrelevant to natural selection.'

Equally puzzling is the way in which some organisms abandon their alleged advantages. If learning to fly provided birds with new opportunities, why have some birds given up flying? Notably the penguin, which seems to be trying to resume its ancestral life as a fish – and not unsuccessfully. The Darwinian thesis of elimination by competition simply cannot be correct as it stands.

Or take mimicry. We are told that butterflies escape their predators by acquiring a resemblance to distasteful butterflies which birds have learned to avoid. Would it not have been simpler to acquire distastefulness? But, in any case, we have to explain why the Black Archer butterfly which makes no concessions to avoiding recognition is so successful that it sometimes strips entire forests. Conversely, if avoiding identification is advantageous, why are there few green butterflies? Green pigment is plentifully available in the caterpillar stage. The elaborate painting of butterfly wings seems an invitation to predators, and the movement of these wings even when the creature is at rest on a plant makes it even more likely that a predator will spot it.

A third phenomenon which gives one pause is the way evolution achieves the same end by different means – not only in different organisms but sometimes in the same organism. I have mentioned how a damaged eye may be reconstructed using tissues different from those from which it was originally derived.

It is this which has given so many biologists the feeling that there must be a plan to which development adheres. (We find the same thing in behaviour, of course. If the goal cannot be attained by one manoeuvre, the animal tries another.)

The remarkable absence of intermediate forms is also worrying. Again and again we find the fossil record bare, and then a wealth of forms. Were they really all derived from a common ancestor by intermediates which have all been lost? This is as true of plants as of animals. 'An abominable mystery', is how Darwin summed up the problem of the origin of flowering plants. They first appear in the Cretaceous period, already much diversified, suggesting an origin much earlier. But not a trace of any intermediate stage has been found. We do not even know whether they evolved from the gymnosperms – the conifers – or whether they were a separate development.

Another unsatisfactory situation is the unnecessary elaboration of many adaptations. The cow, for instance, has a complicated double stomach for digesting grass. It would be easy to say that this stomach evolved because it was more efficient. But horses have a simple stomach and manage very well, so is it true that the rumen is really an advantage? The lady's slipper orchid has an immensely complicated system of fertilisation – and is on the verge of extinction.

A problem of a more philosophical nature is presented by the continual increase in complexity as we ascend the evolutionary scale – sometimes referred to as anagenesis. If mere survival is the criterion of success, what was wrong with the rabbit, as Professor von Bertalanffy, of the State University of New York, asks. Or for that matter with bacteria?

Finally, there is the mystery of behaviour. The fact that organisms, or at any rate animals, *choose* their habitats and can flee from inhospitable environments, pulls the rug from under the whole Darwinian hypothesis of natural selection. There are, of course, many lesser problems in evolution which I have not mentioned. One is the role played by sex: it is not at all clear when sexual reproduction is advantageous and some organisms manage very well without it, while others enjoy both sexual and asexual reproduction according to need.

The origin of the metazoa remains a complete mystery. And I could add to the list the phenotype problem. All in all, evolution remains almost as much of a puzzle as it was before Darwin advanced his thesis. Natural selection explains a small part of what occurs: the bulk remains unexplained.

Darwinism is not so much a theory, as a sub-section of some theory as yet unformulated. Its greatest weakness is the fact that it cannot be disproved. For every circumstance it is possible to imagine some justification. If a species survives, we are told that it adapted. If it fails to survive, that it failed to adapt. If it displays some unusual feature, we are told that it is of adaptive advantage – but whether it really is or not can never be proved. 'I for one . . . am still at a loss to know why it is of selective advantage for the eels of Comacchio to travel perilously to the Sargasso sea, or why Ascaris [a parasitic worm which infests man] has to migrate all round the host's body instead of comfortably settling in the intestine where it belongs . . .' complains von Bertalanffy. (Ascaris has a life-cycle which is relatively simple, as the lives of parasites go. The young larvae enter the host's intestine with the food he consumes. They pass through the intestinal wall into the blood stream, and so via the portal vein to the liver, and from here into the lung and throat. Here they must again be swallowed, so that they arrive back, now in a state of sexual maturity, in the intestine, from which they are eventually excreted.)

'I think the fact that a theory so vague, so insufficiently verifiable . . . has become a dogma can only be explained on sociological grounds,' von Bertalanffy concludes.

Some much more general theory seems to be required if a comprehensive account of evolution is to be written, involving principles of organisation which transcend those of genetic control.

In saying this I do not wish to denigrate Darwin, whose understanding of the difficulties in his theory went far beyond that of many of his followers. He was by any standards a genius, and it is no derogation that, a century later, his ideas are seen to need a larger context than he could have hoped to give them. The synthetic theory is one of the most impressive achievements of the human intellect. That Darwin's ideas need setting in a larger framework no more detracts from his greatness than Einstein's theory did from Newton's.

But now I want to turn to the vexed question of Lamarckism. It is of central importance, because Lamarck stressed the evolutionary importance of behaviour.

2 *Is Evolution Darwinian?*

The core of neo-Darwinism is that variation arises in the genome, irrespective of environmental conditions, which simply select or reject the variations. Information, in short, flows out of the genome and never in. The core of Lamarckism is the exact contrary: that information flows into the genome from the environment and only in.

Scientists relish such dichotomies but the event usually shows that they are false. In reality both factors are interwoven.

Philosophically, it seems most unlikely that the Darwinian view could be correct, (and even more unlikely that the Lamarckian view could be). If organisms need to adjust to the environment, it is logical that they should be influenced by it. The advantages of being able to respond to circumstances are so great that natural selection might be expected to favour very strongly any incipient mechanism with this effect. Such response would of course have to be sluggish. If the organism tried to respond to every minor, short-term change, it would constantly be adapted to yesterday's situation.

For Professor Rupert Riedl, of the Zoology Department of Vienna University, who has made a broad study of order in living organisms, the idea that a creature should be put in an environment to which it has to adapt but be totally deprived of information about that environment, is 'preposterous and indeed unbelievable and catastrophic'. That the creature should achieve the correct response by chance mutation is 'as unlikely as the enhancement of a good poem by a printer's error'.

As we have seen, the arguments advanced by the Darwinians against any trace of inflowing information are without validity. The fact that Lamarckian inheritance has not been shown experimentally – in experiments running for fifty or so generations – proves nothing, since the 'sluggishness' requirement could mean that 10,000 generations are required. It could be said with equal justice that the appearance of a new genus has never been shown experimentally either. It is a mark of the arrogance of Darwinians that they have managed to throw the onus of proof on to the Lamarckians.

Their second argument was that information flows only outward from the genome – the so-called Central Dogma. The work of Howard Temin, of the University of Wisconsin, showed this to be untrue. The fact is, the question of Lamarckism remains wide open, despite the almost hysterical claims of Darwinians that it has been settled once and for all.

The critical fact, in my view, is that the environment does alter the phenotype – the skin on the soles of our feet does thicken when we walk barefoot. The ostrich does acquire calluses where it sits. The output of red blood cells does rise when we live at high altitudes. The number of sweat glands does increase when we move to the tropics. In all these cases the genome of body cells has undoubtedly been affected though the mechanism remains totally mysterious. Until we understand it, we are in no position to dogmatise about the inward flow of information.

Let us move from the general to the particular and remind ourselves of the case of the flatfish, in which, like soles and turbot, the eye, which was once underneath, has worked its way round to the other side, so that both eyes are now on what is effectively the top of the animal, i.e. both are on the same side. As we know, the advantage to the fish is clear.

Clearly it is the behaviour of the flatfish which is the prime cause of the morphological change: its 'decision' to sleep on the bottom on its side. Most accounts stop there; the writer thinks he has explained everything. But the really awkward question remains: how did the genome (if it was the genome) know that an eye shift was required? The official doctrine is that the fish had to wait until, by chance, a genetic change occurred. But can there really be a gene for shifting one's eyes about? In point of fact, the genome may not be in charge of morphology at all. But if it is, I find it inconceivable that such a practical change should depend on chance.

There is another point. The sole's eye does not suddenly appear in a new position in the offspring of normally constructed soles, as one would expect if it were the result of a point-mutation or chromosomal rearrangement. It works its way round over uncounted generations – there is even a species in which the eye has just reached the edge, between top and underside. What mechanism is it which works in this way? We simply do not know. Much the same seems to be true of other adaptations of the Lamarckian kind. How the controlling mechanism knows that it is worth persisting in bringing about a change which may, at some future date, be advantageous, is a matter so obscure that biologists do not even offer theories about it, but brush it under the carpet.

In the evolution of mammals, it is emphatically the case that adaptation is not so much to new environments as to new modes of life, as evolutionists are slowly coming to realise.

Darwinians have assumed that Lamarckism necessarily implies a direct action of the environment on the genome, but probability considerations suggest that the action may well be indirect. As Rupert Riedl puts it in his important book *Order in Living Organisms*: 'selection does not work from the outside only'. He continues: 'As with the laws of entropy, the causality defined by the genetic dogma is not broken by organisms, but evaded. Neolamarckism postulates that there is a direct feedback. Neodarwinism postulates that there is no feedback. Both are mistaken. Truth lies in the middle. There is a feedback but it is not direct.'

The error biologists have made is to think in terms of simple chains of cause and effect. But in a system which is equipped with feedbacks, we have systems of decisions which include cause and effect in common. Each part is thus just as much cause as effect.

3 Where Is Evolution Going?

If there is one solitary fact which emerges distinctly from evolutionary studies it is that evolution is not the execution of a consummate overall plan, divine or otherwise. There have been far too many false starts, boss shots and changes of intention for that. It is reckoned that at least 90 per cent of all the species which have ever existed have become extinct. We have only to think of the land animals which go back into the sea, the legged animals which become burrowers, the birds which cease to fly, to realise that. Many forms which are normally regarded as successful variants – such as man – turn out on examination to be very poorly designed. The prevalence of 'slipped discs' is one proof of that. Years ago, the Russian scientist Elie Metchnikoff remarked that he could point to at least 120 features of the human body which could have been much better designed by a good engineer.

I am reminded of the comment of the late H. L. Mencken that he could not believe that the world had been designed by God, or a god, since it had so obviously been designed by committees.

More seriously, the metaphor which occurs to me in thinking of the evolutionary process as a whole is that of the rising tide flooding into a broad river estuary. As it fills up the channels and available depressions, the sea extends its coverage but this produces no definable pattern. The height of the tide is correlated with the area finally covered but the pattern depends entirely on the lie of the land. To predict it we don't need further information from the water, we just need a very good orographical map. In the same way, it is the environment which sets constraints to the pattern of evolution. Improved adaptability may open up new niches (as the ability to live in space is doing at the moment) but the niches have to be there.

Nevertheless, even if evolution does not express the execution of a plan, it may still have an identifiable trend. As mathematicians would say, it may be vectored – vector being a term indicating direction without any hint of what is moving, or how smoothly or how fast.

In the eighteenth century, of course, it was taken for granted that man was the summit of evolutionary achievement. (Aristotle, who launched the idea of the Ladder of Nature, went much further, starting with inanimate objects and going beyond man to divine beings.) Since then there has been a massive change of public attitude, tending to deflate man's pretensions. The gradual realisation that the earth was only a minor planet, circling a minor sun, in the suburban areas of a not particularly interesting galaxy did much to bring home man's insignificance on the cosmic scale and it was, I think, part of this general trend that people leaped with iconoclastic enthusiasm on the idea that perhaps he was not even at the summit of the evolutionary tree. The tapeworm, it was gleefully pointed out, has man working to support him, so is not the tapeworm entitled to look down on man as less adapted than itself?

The mistake here, and it is often made, is to assume that adaptedness is the same as evolvedness. Flexibility, not adaptedness, is the name of the game in evolution. We have earlier seen how the highly adapted forms tend to be extinguished, while the more flexible ones survive. Man can adapt: he is adapted to live in the lethal environment of space when he dons a space suit. His plus is that he can take that space suit off when he requires to live in a different environment, whereas the tapeworm cannot survive outside its niche.

However, evolution does not seem to have been evolution in flexibility, even if it has been, as Professor Piaget claims, evolution in ability to exploit the environment. What it does seem to have been (if nothing else) is evolution in complexity. The metazoan is palpably more complex than the protozoan; the eukaryotic cell is strikingly more complex than the prokaryotic cell. As we go up the scale, the differences become less marked but they are there. A warm-blooded animal is more complex, physiologically, than a cold-blooded one. Since man has lost the capacity, possessed by most simians, of manufacturing Vitamin C, the iconoclasts have argued that he cannot claim to be the most complex organism and so, once again, cannot regard himself as king of the castle. But to say that complexity is the criterion cannot possibly mean that every single feature has to be more complex. One must take an overall view. A motor-rally competitor who strips some accessories from his car while at the same time installing electronically controlled ignition has not thereby made it less sophisticated. Since man's brain is by many orders of magnitude more complex than that of apes, the fact that he has to get his Vitamin C by eating pills, oranges and so forth, is trivial.

But is growth in complexity the object of the operation or simply an inescapable side-effect? A modern aircraft is more complex than one built fifty years ago; but not because aircraft designers make complexity their goal. If evolution has a trend it is not one of philosophical interest: as with the rising tide, it is the product of its own built-in need to persist. As has cynically been said, evolution is merely the attempt to perpetuate a base-sequence.

But we need not close the story on this discouraging note, for there are other lines of approach to the problem.

Darwin put ability to survive as the prime criterion of evolutionary success. But something more than survival seems to be at work. Nothing in Darwin predicts a continuous increase in complexity.

4 Order From Disorder

When a cell manufactures a protein, a chain of amino-acids reels off the ribosome – but it does not long remain a simple, linear chain. It promptly folds into a highly complex shape which varies according to the particular sequence of amino-acids incorporated. And it is this shape which endows it with the particular biochemical properties, such as enzymatic action, which it then displays. For instance, in the case of haemoglobin (which I take because it is best understood) the molecule contains just the right-shaped slot to embrace the requisite iron atom. Many enzymes seem to act by fitting themselves to the substance on which they act and slightly distorting it, so that it, in turn, will lock into a third substance. Briefly, the power of enzymes to catalyse reactions depends on their precise shape.

The configuration which the chain takes up is determined by forces between various pairs of atoms: there are electrical and other short-range forces at the atomic level which serve this purpose.

Cells manufacture more than a thousand different enzymes, to say nothing of other proteins such as haemoglobin, each with its special configuration. In short, the biochemistry of the cell is made possible and is prosecuted because things are the way they are. Amino-acids are such and the forces they exert on each other are such that the complex tools of metabolism assemble themselves automatically. This is such an extraordinary fact that its importance is overlooked or taken for granted all too often.

Nor is self-assembly confined to amino-acid chains. For further example, the little reels of histone upon which exactly two-and-a-half turns of DNA are wound start life as discs which show a propensity to stack up in just the required way. Similarly, the tubules which the electron microscope reveals in the fine structure of protoplasm assemble themselves. The raw material of flagella is flagellin. Placed in a dish it rapidly forms itself into isolated flagella. Collagen, the gristly stuff which separates our spinal discs (among other functions) assembles itself readily. So do ribosomes, more surprisingly. Many kinds of membrane are self-assembling, and there is some evidence that mitochondria may be.

It seems, then, that in the biological sphere at least, there is a built-in tendency to self-assembly at the most elementary level. That there is also such a tendency at higher levels is obvious: the development of the fertilised egg into an adult organism is a standing example.

In the non-living world, though atoms assemble themselves into molecules, we do not find molecules assembling themselves, except in the one instance of crystal growth. In general, crystals consist of a regular array of molecules, like bricks in a wall; the only new properties which result from such arrays are unusual ways of refracting light, X-rays, and other vibrations, and the harmonising of such vibrations as manifested in the laser. In contrast, living material may be regarded as an irregular, or aperiodic, crystal – as Ernst Schrödinger, the great atomic physicist, suggested nearly forty years ago.

(An intriguing exception to the above generalisation is the snowflake, in which water molecules crystallise in scores of variant shapes.'

Professor Erich Jantsch sought to extend the idea of self-assembly to the whole cosmos in his last book, *The Self-Organizing Universe*. And of course it is true that, on the one hand, we have stars and black holes and quasars just because the properties of matter are such as to produce them; and on the other that in society, or assemblages of living creatures, a structure emerges. Pecking orders are devised, territories staked out – and in the case of man institutions are set up, between which transactions occur in a manner rather analogous to what happens in the developing cell.

However, the universality of self-assembly need not be pressed. The point which is relevant here – more than relevant, crucial, I should say – is that the forms we see in the course of evolution are what they are not only because of environmental constraints but also because of built-in necessity. After I had reached this conclusion I realised that an old friend, the late Lancelot Law Whyte, had in fact provided me with the answer in 1965 when he sent me a proof of his highly original little book *Internal Factors in Evolution*. (Whyte, while still at Cambridge, became convinced that physicists grossly exaggerated the degree of disorder in the universe and spent his life studying the causes of order.)

Whyte's idea was that the genome is self-stabilising; it will only accept mutations which improve or at least do not disturb its stability. In other words: 'Only those changes which result in a mutated system that satisfies certain stringent physical, chemical, and functional conditions will be able to survive the complex chromosomal, nuclear and cellular activities involved in the processes of cell division, growth and function.' Thus the number of possible variations is limited. Possibly the genome can modify mutations which are nearly acceptable into full acceptability. Probably it can handle groups of mutations, each of which alone might be unacceptable, if the overall effect is stabilising.

If Whyte is right, no mutation is entirely due to chance: only those which meet the internal demands of the genome can emerge to be selected by Darwinian processes. In fact, many people before Whyte have wondered if some inner principle was at work, from the German biologist von Baer (famous as the man who discovered the mammalian egg) in 1876, onward. Perhaps the first to define it as internal selection was General Smuts in 1926; since World War II there have been at least a dozen.

Whyte believed that this idea was probably one of the most fertile biological ideas of the century and was puzzled that it had not received wider recognition. I think the fact that it does not at present point to practicable experiments may be the reason it has been neglected, for it is by mounting research projects that scientists earn their living. There are plenty of unanswered questions, to be sure, and Whyte lists a selection of them, but we need to know more about development before we can attempt to elucidate them.

I have told how sea urchins fall apart in the absence of calcium, but reassemble themselves in their old shape if calcium is supplied. Here is a vivid instance of the built-in necessity to achieve form, to produce order out of chaos. I can give you many others. The division of the cells in a plant leaf seems random, but from it emerges the organised structure of the leaf. Nothing is more chaotic than the tangled hyphae, or feeding tubes, put out by fungi. Yet from them arises, by means unknown, the organised 'fruiting body' which propagates such organisms.

That order should arise spontaneously from disorder is a fact so astonishing as to deserve further discussion. It may seem, at first, to contravene the second law of thermodynamics, but this is not so. That law says that heat flows from hotter areas to cooler ones, so that eventually all ends up at an even temperature. It is often assumed that this also means that order becomes disorder. However, there are highly organised crystals which maintain their structure at absolute zero. Conversely, at high temperatures, the atoms in a gas are in a chaotic state. The two propostions are unconnected. It is true that ordered structures tend to break down and much energy must be spent and ingenuity exercised in reassembling them. A broken egg is not easily reconstituted. It is the peculiarity of life that it produces such order. Certainly, it does so within the limits of entropy, and derives its energy from the ultimate flow of heat from hot to cold; that is, from the sun. But that does not alter the fact. Order is often increased in open systems into which energy flows. The problem is the increase in information content, i.e. in the level of organisation.

The subject of how order arises in such systems had been studied by physicists for many years and in recent decades they have evolved a new branch of the subject capable of handling these non-equilibrium systems. First firmly stated by I. Prigogine of the Free University of Brussels in 1937 and later applied specifically to biological systems by Manfred Eigen of Frankfurt University, it is known as Non-equilibrium Thermodynamics or the thermodynamics of irreversible processes.

The topic is complicated, but one point which concerns us stands out: as any system grows in size and complexity, it becomes unstable. It may collapse. Living systems, however, usually reorganise themselves and may do so quite abruptly. They flip. This has been termed 'progress by fluctuation'. The change of structure which occurred when the invertebrates gave rise to the vertebrates may be a dramatic instance of such a fluctuation or flip; but the story of evolution bristles with such sudden reorganisations, as we have seen. And they are irreversible.

The theory states that a steady throughput and a steady selection of more stable conditions necessarily causes a steady increase in order. The result is a self-ordering of living systems, amounting to self-design.

What is perhaps less easily understood is the tendency which some computer programmers have found for order to emerge from disorder in computer simulations. Stewart Kauffman at the University of Pennsylvania is one of these programmers. When he programmes his computer to resemble a block of cells, any one of which can stimulate any neighbouring cell, he gets, not a random distribution of stimulated cells, but recognisable patterns.

A gene system with 10,000 genes could produce 10^{-3000} possible combinations. But Kauffman's system, based on some simple assumptions about the way in which interactions occur, spontaneously confines itself to about one hundred different patterns of gene action. He proposes that these correspond to the hundred or so different types of cell found in complex organisms. 'There are powerful deep underlying constraints that do not need natural selection.'

Organisms, it is beginning to be realised, are not pieces of putty but resilient structures with their own internal necessities and potentialities.

More unsettling still is the discovery by computer simulations that the whole process of speciation, on which so much intellectual labour has been expended, may in reality be the product of such nominally random forces. Professor David Raup at the University of Rochester devised a computer simulation of the speciation process. He set up, on his computer, a number of species, the only condition being that they could do one of three things in any given time period: persist, branch into two daughter species, or become extinct, on a random basis.

When the programme was run, the simulated species behaved almost exactly as in the real world. Some proliferated rapidly and then became extinct rapidly, just as the trilobites did. Others increased slowly but vanished rapidly, as the dinosaurs did. Yet others had several peaks of success, like the dinosaurs. Sometimes one species replaced another, as if there had been competition. (However, there were no mass extinctions and no exceptional long survivals, corresponding to living fossils, such as the coelacanth.)

There were further similarities. At the start there had been eleven lines of descent; at the end there were only four, with gaps between them just as there are among mammals. In some cases there were 'trends' (orthogenesis). There was also 'parallel evolution'. However, 'convergent evolution' was not found. Raup concluded that the evolutionary pattern as we know it could quite well have been generated, not by natural selection, but by purely random events.

Professor Douglas Futuyama, of the State University of New York at Stony Brook, after examining these results, calls them 'quite unsettling'.

5 A Matter of Form

All biology reduces to two elements: substance and form. That is, the materials of which an organism is made and the manner in which those materials are arranged. During the last half century, biologists have discovered a great deal about substance. We understand, in a large measure, what proteins, fats and fibres make up the cell, and also how they are manufactured. We remain extremely ignorant about how these raw materials are arranged in the complex structures we observe.

Many authorities have stressed the importance of form in biology. The late Professor C. H. Waddington went as far as to say: 'The whole science of biology has its origin in the study of form.' More tersely, Professor Joseph Needham (nowadays known to most people as the historian of Chinese science but originally an embryologist) says: 'The central problem of biology is the form problem.' Alas, it is an unsolved problem and the Australian biologist, W. E. Agar, who has studied it, concludes that how the organism is produced from relatively formless components is 'the great enigma of biology'.

But it has been little studied. D'Arcy Wentworth Thompson's classic work *On Growth and Form*, first published in 1917, is the only major contribution to the discussion of the topic, with the honourable exception of some books by a Yale professor, Edmund Sinnott, which have not received the attention they deserve.

Too many biologists assume that, somehow, the genetic material prescribes the form of the organism as well as specifying the raw materials. This cannot be entirely true and is probably not true at all. If the little worm *Naegleria* is placed in a strong salt solution it develops flagella; in a weak one it does not. So form is certainly dependent on external factors, at least in some cases.

Howard Pattee, of the Biophysics Laboratory at Stanford University, sticking closely to the facts of molecular biology, points out that the degree of order found in biological macromolecules cannot be adequately explained on the basis of genetic controls alone, acquired by natural selection from primeval sequences. There are several strange facts. Thus, on the one hand, several proteins perform similar functions though they vary widely in structure; also, well-ordered sequences of amino-acids can be removed from proteins without affecting their function. (So why are they ordered?) Again, it is odd that the distribution of amino-acids does not follow any noticeable order. He infers the possibility that the precursors of biological molecules were not random sequences but were 'naturally-ordered crystal structures which resulted from the restrictions inherent in their growth'. Indeed, he goes further when he draws the inference that the statistical evidence from amino-acid sequences is consistent with some degree of *non-genetic* ordering.

Paul Weiss, too, of Rockefeller University, New York, has argued that genes account for structural differences but not for organisational ones. 'There is neither logical nor factual support for the supposition that organisations can be explained in reference to "gene",' he writes. 'This claim rests on sheer assertion, based on blind faith and unqualified reductionist preconceptions.' Strong words!

Since change of form is the core of evolutionary studies, if Weiss is right, the picture we have lacks the most important component of all. We cannot say that evolution has been 'explained' until we know more about the factors which determine form.

One of the mysterious features of biological form is why there are so many – far more than is necessary. D'Arcy Wentworth Thompson found this multiplicity of forms actually inconsistent with the idea of natural selection. He thought it particularly strange that the same forms so often recur in widely varying conditions. 'We find the same forms which (save for external ornament) are mathematically identical repeating themselves in all periods of the world's geological history: and we see them mixed up, one with another, irrespective of climate or local conditions, in the depths and on the shores of every sea. It is hard indeed, to my mind, to see in such a case as this where natural selection enters in, or to admit that it has had any share whatever in the production of these varied conformations.'

When we see the restoration of a damaged part, often by the use of materials different from those used initially, we have such a strong impression of the existence of a plan, dictating form, that some biologists and many laymen have been convinced of the existence of a plan or purpose. On the other hand, many aspects of form seem quite arbitrary. Why are the leaves of some plants placed in pairs, and others alternately? Still more pointless, it would seem, is the way in which the leaves of common ivy vary between the vegetative and the flowering shoots.

It has been surmised that people tend to fall into one of two personality categories: substance-minded and form-minded. The former, it is said, tend to concentrate on practical issues, on getting things done. The latter are more concerned with general principles and the philosophy underlying action. If there is anything in this, biologists would seem to be drawn mainly from the former group. I am struck by a medical parallel. Until recently, the trend when seeking the cause of a disease, has been to look for pathogenic agents, such as microbes and viruses – in a word, for substances. Today, however, we look increasingly for disturbances in regulation and control, which are central in cancer, auto-immune disease and so on.

There must be 'laws of form' but unhappily we do not know what they are. Recently, an idea has been put forward which is of particular relevance to evolution and which ties in with the work of Manfred Eigen which I have mentioned. It is that there are certain 'decision points' in the growth of an organism where one of two routes must be taken, and there is no going back. For instance, if a flat sheet of cells multiplies in such a way that the centre is increasing in size more rapidly than the periphery, a pouch or bulge is bound to form. Then, either this pouch goes *in* (as it does in the growing sphere of cells which constitutes an early embryo) or it must stick *out*, like a thumb.

In the course of evolution we can perceive many such decision points, such as the 'decision' to have four legs in amphibia or six in insects. Or again, the switch from radial symmetry to bilateral symmetry which took place when the vertebrates first appeared. (Such branch points also occur in the growth of social organisations, such as cities. The siting of a large town may exercise an irreversible effect on the siting of surrounding villages and the traffic flow between them.) Some mechanism of this kind, therefore, may explain the distinctions between the phyla, and prove gradualism wrong. The forms of order may be identical in civilised and in somatic evolution.

Until we understand the laws of form we are in no position to say that we understand the mechanism of evolution.

Progress has been prevented by the rigid dogmatism of the Neo-Darwinians and it is quite significant that the first crack in the structure they erected has come on the subject of 'punctuation'. That is, on the question: is evolution gradual or do abrupt discontinuities occur? As Professor Rupert Riedl of Vienna puts it: 'Most of the unexplained phenomena in macro-evolution were first minimised, then swept under the carpet and finally forgotten.'

Punctuationism is significant in that once it is conceded that some other mechanism than natural selection operates in evolution, even if only from time to time, natural selection is ousted from its unique position and Darwin's idea becomes merely a part of a larger theory. Of course, once one admission of this calibre is made, the situation becomes much more fluid and scientists will feel free to look around for new interpretations in a much less inhibited manner, free from any fear of having their careers damaged by the awful charge of unorthodoxy.

Actually, the origin of the phyla is not by any means the weakest point in the Darwinian position. Many facts remain inexplicable, as we have seen. Modern biology is challenged by 'a whole group of problems' as Riedl remarks. Now, however, the attempt to present Darwinism as an established dogma, immune from criticism, is disintegrating. At last the intellectual log-jam is breaking up. So we may be on the verge of major advances. The years ahead could be exciting. Many of these advances, I confidently predict, will be concerned with form.

It is unfortunate that the Creationists are exploiting this new atmosphere by pressing their position; this naturally drives the biologists into defensive attitudes and discourages them from making any admissions.

Evolutionists have been blinkered by a too narrowly materialist and reductionist approach to their problems. But the trend of the times is away from Victorian certainties and Edwardian rigidities. In the world as a whole, there is a growing recognition that life is more complex, even more mysterious, than we have supposed. The probability that some things will never be understood no longer seems so frightening as it did. The probability that there are forces at work in the universe of which we have as yet scarcely an inkling is not too bizarre to entertain. This is a step towards the freeing of the human mind which is pregnant with promise.

Glossary

Allopatric speciation The evolution of two or more distinct species from geographically isolated populations of a single species

Amino-acid A compound containing an amino (NH_2) and a carboxyl (COOH) group. There are twenty amino acids which, in various combinations, give rise to all known proteins

Amniote egg The egg of a land-living vertebrate (reptile, bird or mammal) whose embryos develop in a fluid-filled sac or amnion

Anaerobic Lacking in oxygen

Anagenesis The regeneration of tissues

Angiosperm Flowering plant

Aplanatic lens Lens possessing no spherical aberration

Aristogenesis Term referring to the process by which the germ plasm (hereditary material) gives rise to new bio-mechanisms

Basilar membrane Thin vibratile membrane underlying the organ of Corti in the cochlea of the inner ear

BP Before the present – used in referring to the age of fossils

Carpel Female parts of a flower consisting of an ovary, a style and a stigma

Chert Sedimentary siliceous deposit which has been lithified (turned into stone) by redistributed silica

Chitin A major component of arthropod exoskeletons (external skeletons)

Chlorophyll Green pigment which makes it possible for plants to synthesise their food (see **Photosynthesis**)

Chordate An animal having a notochord (later replaced by the skull and backbone in vertebrates), a hollow nerve tube and gill slits at some stage in embryonic development

Chromosome Body containing the hereditary material found in the nuclei of living cells. Each species has a constant number of chromosomes

Cichlid Family of freshwater fishes found principally in Central America, tropical parts of South America, and in Africa. Some cichlids, such as *Tilapia*, are important food fishes

Cilium (pl. cilia) Mobile hair or bristle-like outgrowth from the surface of cells

Cladistics A method of classifying organisms according to their 'recency of common descent'

Cleidoic egg Egg possessing a shell which is only permeable to gases

Co-arcevate Cluster of electrically charged protein or protein-like molecules which form themselves into a small 'droplet' surrounded by a layer of water molecules

Convergent evolution The evolution of unrelated organisms which occupy similar niches and which bear superficially similar characteristics

Cope's Law 'Law' formulated by the paleontologist E. D. Cope in the nineteenth century stating that, in the course of evolution, taxonomic groups showed a tendency to evolve towards a progressively greater size

Cytoplasm The substance within a cell, excluding that found within the nucleus

Darwinism Theory of evolution formulated by Charles Darwin based on the theory of natural selection

DNA Deoxyribonucleic acid: molecule carrying hereditary information in the form of a transversely linked double helix

Endosymbiosis 'Internal' symbiosis in which two or more organisms of different species coexist in an intimate and mutually beneficial partnership: e.g., the algae which live in the cells of some species of *Hydra* and corals

Entropy The degree of randomness within a molecular system or the amount of unavailable energy in a thermodynamic system

Enzyme A proteinaceous biological catalyst which acts on one or a few substrates only

Ethology The scientific study of behaviour

Eukaryote An organism made up of cells which contain a true nucleus, i.e., one bounded by a nuclear membrane

Flagellum (pl. flagella) Long, thread-like outgrowth from the surface of cells such as gametes (reproductive cells) and single-celled organisms. Flagella are responsible for the swimming movements of such cells

Frame shift mutation A hereditary change in which a base (such as guanine) is either inserted into or removed from the information carried by a gene

Gene The basic hereditary unit of living organisms

Genome The complete set of genetic information contained in a gamete (reproductive cell)

Genotype The genetic constitution of an organism as opposed to its physical constitution (phenotype)

Golgi apparatus System of cavities found in the cytoplasm of cells concerned with the synthesis and secretion of a variety of substances

Gradualism Theory that evolution proceeds by small 'steps' over a long period of time

Groeberiidae Family of extinct rodent-like marsupials (pouch-bearing mammals)

Guanine One of the bases found in nucleic acids such as DNA, the others in DNA being adenine, cytosine and thymine

Gymnosperm A division of the plant kingdom consisting of conifers and related plants

Histone A protein of high molecular weight associated with DNA

Homology Similarity of structures resulting from common ancestry. Homologous structures do not necessarily perform similar functions (e.g.) the ear ossicles (bones) of mammals are homologous with some of the bones involved in jaw attachment in fishes

Ichthyology The scientific study of fishes

Invertebrate Animal without a backbone

Lamarckism Evolutionary theory put forward by Jean-Baptiste Lamarck in 1809 based on the inheritance of acquired characteristics

Macroevolution Large-scale evolution (i.e.) of genera or species through mutations which produce marked changes over periods of geological time

Melanism The genetically controlled development of an excessive amount of black pigment

Mendelism Rules formulated by Gregor Mendel governing the inheritance of characters according to predictable ratios

Metazoa Animals whose bodies are made up of numerous cells

Microevolution Small-scale evolution: e.g., of individual genes, producing relatively minor changes over a short period of time, such as a few generations

Mitochondrion A self-replicating organelle (small cytoplasmic structure) involved in energy production

Morphogenesis The progressive development of organs or other parts of organisms culminating in the characteristic adult form

Morphology The study of the form and structure of organisms

Neo-Darwinism Evolutionary theory combining the theory of natural selection with Mendelism

Ontogeny The development of an individual from the egg to the adult form

Organelle Structured parts of a cell's cytoplasm responsible for specific functions

Orthogenesis Evolution in a particular pre-determined direction not influenced by natural selection

Oscillatoria A genus of filamentous blue-green algae

Paedogenesis Reproduction in pre-adult or larval stages of an organism: e.g., as in axolotls

Parallel evolution The independent evolution of similar characteristics in organisms from related evolutionary lines

Parapatric speciation The evolution of species in non-overlapping geographical regions which are nevertheless in contact

Parthenogenesis Reproduction in which no fertilisation occurs

Phacopid A mid-Cambrian to Devonian line of trilobites having relatively large 'advanced' eyes, the visual parts of which were attached to the cheek

Phene A genetically controlled physical characteristic

Phenotype The physical constitution/appearance of an organism resulting from the interaction between the genotype and the environment

Pheromone A substance secreted by one animal that affects the behaviour of another of the same species or a closely related one

Photosynthesis Process by which chlorophyll-bearing plants synthesise food, in the form of carbohydrate, from carbon dioxide and water in the presence of light; a by-product of the process is oxygen

Phylogeny Study of the evolutionary history of a group of organisms

Placental Possessing a placenta (as in mammals)

Polydactyly Possession of a number of digits in excess of the normal

Polymorphism The occurrence of two or more genetic forms of a single species

Prebiotic Before the existence of life

Prokaryote An organism lacking a well-defined nucleus: e.g., bacteria, blue-green algae

Protein Complex molecule made up of amino-acids

Protoplasm The substance within a cell, including the nucleus

Prototherian Egg-laying mammal

Protozoa Single-celled (unicellular) animals

Pseudosuchians Extinct reptiles generally believed to have given rise to the birds

Punctuation theory Theory that evolution proceeds in a series of 'jumps' with periods of relative stability in between

RNA Ribonucleic acid; molecule containing the bases adenine, guanine, cytosine and uracil and involved in protein synthesis

Ribosome Minute granules, very rich in RNA and playing a vital role in protein synthesis

Saltation — Abrupt change brought about by a major mutation

Silicate — A chemical substance whose basic units consist of ions of silica and oxygen in the ratio of 1:4 in a tetrahedral arrangement

Somatic — Pertaining to the bodily (as opposed to the germinal) part of an organism

Speciation — The formation of two or more new species from a single existing one

Spirochaetes — Spiral bacteria lacking flagella but capable of movement by undulations of the body

Symbiosis — A mutually beneficial close association between two unrelated organisms

Sympatric speciation — The evolution of species without geographical isolation

Taxon (pl. taxona) — Any group of organisms which is sufficiently distinct from all others to be accorded a scientific name: e.g., species, family, class, etc.

Taxonomy — The study of the classification of organisms

Therapsid — Extinct mammal-like reptile

Vertebrate — Animal having a skull and backbone

Zeuglodont — Extinct group of aquatic mammals known as 'yoke-toothed' whales

Bibliography

ABLESON, J., 'A revolution in biology', *Science* (1980), *209*:1319

ALLEN, P. M., 'Evolution, population dynamics and stability', *Proc. Nat. Acad. Sc. USA* (1976), *73*(1)665

AMOS, W. H., 'The life of a sand dune', *Sc. Amer.* (1959), *201*(1)91

ANDREWS, S. M., *et al.* (eds.), *Problems in Vertebrate Evolution*, Academic Press, London and New York, 1977

ARDITTI, J., 'Orchids', *Sc. Amer.* (1966), *214*(1)70

ARNON, D. I., 'The role of light in photosynthesis', *Sc. Amer.* (1960), *203*(5)104

ASHWORTH, J. M., *Cell Differentiation*, Chapman and Hall, London, 1973

AYALA, F. J., 'The mechanisms of evolution', *Sc. Amer.* (1978), *239*(3)48

BAERENDS, G., *et. al.* (eds.), *Function and Evolution in Behaviour*, Oxford University Press, 1976

BAKKER, R. T., 'Dinosaur feeding behaviour and the origin of flowering plants', *Nature* (1978), *274*:661

BANKS, H. P., 'Major evolutionary events', *Biol. Rev.* (1970), *47*:451

BARGHOORN, E., 'The oldest fossils', *Sc. Amer.* (1971), *224*(5)30

BARTHOLOMEW, G. A., and HUDSON, J. W., 'Desert ground squirrels', *Sc. Amer.* (1961), *205*(5)107

BATTERSBY, A., *et al.*, 'Biosynthesis of the pigments of life: formation of the macrocycle', *Nature* (1980), *285*:17

BECK, C. B. (ed.), *Origin and Early Evolution of Angiosperms*, Columbia University Press, 1976

BENNET, A. F., and RUBEN, J. A., 'Endothermy and activity in vertebrates', *Science* (1979), *206*:649

BERG, H. C., 'How bacteria swim', *Sc. Amer.* (1975), *233*(2)36

BERMANT, G. (ed.), *Perspectives on Animal Behavior: a first course*, Scott, Foresman & Co, Glenview, Ill., and Brighton, UK

BERRILL, N. J., *The Origin of Vertebrates*, The Clarendon Press, Oxford, 1955

BERTALANFFY, L. von, *The Problems of Life*, Watts, New York, 1952
— 'Chance or law', in Koestler, *Beyond Reductionism*, p.56
BISHOP, J. A., and COOK, L. M., 'Moths, melanism and clean air', *Sc. Amer.* (1975), *232*(1)90
BLOESER, B., *et al.*, 'Chitinozoans from the late Precambrian Chuar group of the Grand Canyon, Arizona', *Science* (1977), *195*:676
BLUM, H. F., *Time's Arrow and Evolution*, Princeton University Press, 1968
BOARDMAN, R. S., *et al.* (eds.), *Animal Colonies*, Dowden, Hutchinson and Ross, Stroudsburg, Pa., 1973
BODMER, W., 'Evolutionary significance of HL-A systems', *Nature* (1972), *237*: 139
— and SFORZA, C., *Genetics, Evolution and Man*, Freeman, San Francisco, 1976
BONNER, James, *The Molecular Biology of Development*, Clarendon Press, Oxford, 1965
BONNER, J. T., *On Development*, Harvard University Press, 1974
— 'Differentiation in social amoebae', *Sc. Amer.* (1959), *201*(6)152
— 'How slime molds communicate', *Sc. Amer.* (1963), *209*(2)84
BRITTEN, R., and DAVIDSON, E. H., 'Gene regulation for higher cells: a theory', *Science* (1969), *165*: 349
— 'Repetitive and non-repetitive sequences and a speculation on the origin of evolutionary novelty', *Qtly Rev. Biol.* (1971), *46*:113
— and KOHNE, D. E., 'Repeated segments of DNA', *Sc. Amer.* (1970), *222*(4)24
BRODA, E., *The Evolution of the Bioenergetic Processes*, Pergamon Press, Oxford, 1975
BROOKS, J., and SHAW, G., *Origin and Development of Living Systems*, Academic Press, London and New York, 1973
BROUGH, J., 'Time and evolution', in Westoll (ed.), *Studies on Fossil Vertebrates*, Ch. 2
BRUN, J., 'Genetic adaptation of crenorhabditis elegans to high temperatures', *Science* (1966), *150*:1467
BRYSON, V., and VOGEL, H. J. (eds.), *Evolving Genes and Proteins*: a symposium, Academic Press, London and New York, 1965
BUFFETAUT, E., 'The evolution of the Crocodilians', *Sc. Amer.* (1979), *241*(4)124
BULLOCK, T. H., 'Physiological bases of behaviour', in Moore (ed.), *Ideas in Modern Biology*, p.449
BULLOUGH, W. S., *The Evolution of Differentiation*, Academic Press, London and New York, 1967
BURKHARDT, B. W., *The Spirit of System*, Harvard University Press, 1977
BUSH, G. L., *et al.*, 'Rapid speciation and chromosal evolution in mammals', *Proc. Nat. Acad. Sc. USA* (1977), *74*(9)3942
CALDER, W. A., 'The Kiwi', *Sc. Amer.* (1978), *239*(1)102
CAMMACK, R., *et al.*, 'Ferredoxins: are they living fossils?', *New Scientist* (1971), *51*:696
CANNON, H. G., *The Evolution of Living Things*, Manchester University Press, 1958

— *Lamarck and Modern Genetics*, Manchester University Press, 1959

CAPLAN, A. L. (ed.), *The Sociobiology Debate*, Harper and Row, New York, 1979

CAREY, F. G., 'Fishes with warm bodies', *Sc. Amer.* (1973), *228*(2)36

CARLSON, E. A., *The Gene: a critical history*, W. B. Saunders, Eastbourne, 1966

CARLSON, H. L., 'Chromosome tracers of the origin of species', *Science* (1970), *168*: 1414

CARTER, G. S., *Animal Evolution: a study of recent views of its causes*, Sidgwick and Jackson, London, 1951

CARVER, J. H., 'Prebiotic atmospheric oxygen levels', *Nature* (1981), *292*: 136

CAVALIER-SMITH, T., 'Visualising jumping genes', *Nature* (1977), *270*:10

— 'How selfish is DNA?', *Nature* (1980), *285*:617

CHAMBON, P., 'Split genes', *Sc. Amer.* (1981), *244*(5)48

CHEDD, G., 'The making of a gene', *New Scientist* (1976), *71*:680

CISNE, J. L., 'Trilobites and the origin of arthropods', *Science* (1974), *186*:13

CLARK, R. B., *Dynamics of Metazoan Evolution*, Oxford University Press, 1964

CLARKE, B., 'The causes of biological diversity', *Sc. Amer.* (1975), *233*(2)50

CLARKSON, E. N. K., *Invertebrate Paleontology*, Allen and Unwin, London, 1979

CLEMMEY, H., 'World's oldest animal traces', *Nature* (1976), *261*:577

COHEN, S. N., 'Manipulation of genes', *Sc. Amer.* (1975), *233*(1)24

— 'Transposable genetic elements and plasmid evolution', *Nature* (1976), *263*:731

— and SHAPIRO, J. A., 'Transposable genetic elements', *Sc. Amer.* (1980), *242*(2)36

COLBERT, E. H., *Evolution of the Vertebrates*, Wiley, New York, 1966

CORNER, E. J. H., *The Life of Plants*, Weidenfeld and Nicolson, London, 1964

CREED, R. (ed.), *Ecological Genetics and Evolution*, Blackwell, Oxford, 1971

CRICK, F., 'Central dogma of molecular biology', *Nature* (1970), *227*:561

— 'General model for the chromosomes of higher organisms', *Nature* (1971), *234*:25

— 'Split genes and RNA splicing', *Science* (1979), *204*:264

— and ORGEL, L. E., 'Directed panspermia', *Icarus* (1973), *19*:341

CRONIN, J. E., *et al.*, 'Tempo and mode in hominid evolution', *Nature* (1981), *292*:113

CROW, J. F., 'Genes that violate Mendel's rules', *Sc. Amer.* (1979), *240*(1)104

CROWE, J. N., and COOPER, A. F., Jr, 'Cryptobiosis', *Sc. Amer.* (1971), *225*(6)30

CUDMORE, L. L., *The Center of Life: a Natural History of the Cell*, Quadrangle, New York, 1977; David & Charles, Newton Abbot, 1978

DARLINGTON, C. D., *The Evolution of Genetic Systems*, Oliver & Boyd, Edinburgh, 1958

DARLINGTON, P. J., 'Rates, patterns and effectiveness of evolution in multilevel situations', *Proc. Nat. Acad. Sc. USA* (1976), *73*(4)360

DARNELL, J. E., *et al.*, 'Biogenesis of mRNA: genetic regulation in mammalian cells', *Science* (1973), *181*:1215

DARNELL, J. E., Jr, 'Implications of RNA-RNA splicing in the evolution of eukaryotic cells', *Science* (1978), *202*:1257

DARWIN, C., *The Variation of Animals and Plants under Domestication*, Murray, London, 1875

DAUVILLIER, A., *The Photochemical Origin of Life*, Academic Press, London and New York, 1965

DE BEER, Sir G., *Embryos and Ancestors*, Oxford University Press, 1958 (1st edn. 1940)

— *Atlas of Evolution*, Nelson, London, 1964

DE JONG, W. W., *et al.*, 'Relationship of aardvark to elephants, hyraxes and sea cows from α crystalline sequences', *Nature* (1981), *292*:538

DENTON, E., 'Reflectors and fishes', *Sc. Amer.* (1971), *224*(1)64

DE ROBERTIS, E. M., and GURDON, J. B., 'Gene transplantation and the analysis of development', *Sc. Amer.* (1979), *241*(6)60

DESMOND, A., *The Hot-Blooded Dinosaurs*, Blond and Briggs, London, 1975

DETLEFSEN, S. A., 'The inheritance of acquired characteristics', *Physiol. Rev.* (1925), *5*:144

DICKERSON, R. E., 'Chemical evolution and the origin of life', *Sc. Amer.* (1978), *239*(3)62

— 'Cytochrome C and the evolution of energy metabolism', *Sc. Amer.* (1980), *242*(3)98

DILGER, W., 'The behavior of lovebirds', *Sc. Amer.* (1962), *206*(1)88

DOBZHANSKY, T., *Genetics of the Evolutionary Process*, Columbia University Press, 1970 (4th edn)

— 'Darwinian or oriented evolution?' *Evolution* (1975), *29*:376

DODSON, C. H., *et al.*, 'Biologically active compounds in orchid fragrances', *Science* (1969), *164*:1243

DOOLITTLE, W. F., 'Genes in pieces: were they ever together?', *Nature* (1979), *272*:581

DOSTAL, R., *Integration in Plants*, Harvard University Press, 1967

DOUGHERTY, E. C., ELLSWORTH, C., *et al.*, *The Lower Metazoa* (conference of 1960), University of California Press, 1963

DOVER, G., 'Ignorant DNA', *Nature* (1980), *285*:618

DOWDESWELL, W. H., *The Mechanism of Evolution*, Heinemann, London, 1955 and 1970

DRAKE, E. T. (ed.), *Evolution and Environment*, Yale University Press, 1968

DRAKE, J. W. (ed.), *The Molecular Basis of Mutation*, Holden-Day, San Francisco, 1970

DUBOCHET, J., and NOLL, M., 'Nucleosome arcs and helices', *Science* (1978), *202*:280

DUNCAN, C. J., *Molecular Properties and Evolution of Excitable Cells*, Pergamon Press, Oxford, 1967

EAKIN, R. M., *The Third Eye*, University of California Press, 1973

EDMUNDS, M., *Defence in Animals*, Longmans, London, 1974

EIBL-EIBESFELDT, I., *Ethology: the Biology of Behavior*, Holt, Rinehart & Winston, New York, 1970

EIGEN, M., 'Molecular self-organisation in the early stages of evolution', *Qtly Revs. in Biophys.* (1971), *4*:149

— *et al.*, 'The origin of genetic information', *Sc. Amer.* (1981), *244*(4)78

Eisenberg, J. F. (ed.), *Vertebrate Ecology in the Northern Neotropics*, Smithsonian Inst. Press, Washington DC, 1980

Eldredge N., and Eldredge, M., 'A trilobite odyssey', *Natural History* (1975), *84*:72

— and Gould, S. J., *Phylogenetic Patterns and the Evolutionary Process*, Columbia University Press, 1960

Etkin, W., 'How a tadpole becomes a frog', *Sc. Amer.* (1966), *214*(5)76

— and Gilbert, L. (eds.), *Metamorphosis: a Problem in Developmental Biology*, N. Holland/Appleton-Century-Crofts, New York, 1968

Ewer, R. F., 'Natural selection and neoteny', *A. Biotheoretica* (1960), *13*:161

Fabre, J. H., *The Wonders of Instinct*, Fisher Unwin, London, 1920

Fairbridge, R. W., 'The changing level of the sea', *Sc. Amer.* (1960), *202*(5)70

Fischberg, M., and Blackler, A. W., 'How Cells Specialise', *Sc. Amer.* (1961), *205*(3)124

Floekin, M. (ed.), *Aspects of the Origin of Life*, Pergamon Press, Oxford, 1960

Ford, E. B., *Ecological Genetics*, Methuen, London, 1964

Ford, T. F., 'Life in the Precambrian', *Nature* (1980), *285*:193

Fox, G. E., *et al.*, 'The phylogeny of prokaryotes', *Science* (1980), *209*:457

Frazzetta, T. H., *Complex Adaptations in Evolving Populations*, Sinauer Assocs., Sunderland, Mass., 1975

Fredrick, J. F., and Klein, R. M. (eds.), 'Phylogen and morphogen in algae', *Ann. N.Y. Acad.*, *175*(2)413

Fremlin, J., 'Consciousness and dinosaurs', *New Scientist* (1979), *81*:250

French, V., *et al.*, 'Pattern regulation in epimorphic fields', *Science* (1976), *193*:969

Frieden, E., 'The chemistry of amphibian metamorphosis', *Sc. Amer.* (1963), *209*(5)110

Fryer, G., and Iles, T. D., *The Cichlid Fishes of the Great Lakes of Africa*, Oliver & Boyd, Edinburgh, 1972

— 'Alternative routes to evolutionary success', *Evolution* (1969), *23*:359

Fujii, T., 'Inherited disorders in the regulation of serum calcium in rats raised from parathyroidectomised mothers', *Nature* (1978), *273*:236

Futuyama, D. J., *Evolutionary Biology*, Sinauer Assocs., Sunderland, Mass., 1979

Fyfe, W. S., 'Effects on biological evolution of changes in ocean chemistry', *Nature* (1977), *267*:510

Gally, J. A., and Edelman, G. M., 'Somatic translocation of antibody genes', *Nature* (1970), *227*:34

Gardner, M. R., and Ashby, W. R., 'Connections of large dynamic (cybernetic) systems: critical values for stability', *Nature* (1970), *228*:784

Ghiselin, M. T., *Triumph of the Darwinian Method*, University of California Press, 1973

Gilbert, W., 'Why genes in pieces?' *Nature* (1978), *271*:581

Gilliard, E. T., 'The evolution of bower birds', *Sc. Amer.* (1963), *209*(2)38

Glaessner, M. F., 'Precambrian animals', *Sc. Amer.* (1961), *204*(3)72

GOLDSCHMIDT, R., *The Material Basis of Evolution*, Yale University Press, 1940
GOOD, R., *The Philosophy of Evolution*, Element Books, 1981
GOODENOUGH, U., and LEVINE, R. P., 'The genetic activity of mitochondria and chloroplasts', *Sc. Amer.* (1970), *223*(5)22
GOODMAN, M., *et al.*, 'Molecular evolution in the descent of man', *Nature* (1971), *233*:614
GOODWIN, D., *Pigeons and Doves of the World*, Cornell University Press, 1967
GORCZYNSKI, R., and STEELE, E., 'Inheritance of acquired immunological tolerance to foreign histocompatability antigens', *Proc. Nat. Acad. Sc. USA, 77*:2871
GORINI, L., 'Antibiotics and the genetic code', *Sc. Amer.* (1966), *214*(4)102
GOULD, S. J., *Ontogeny and Phylogeny*, Harvard University Press, 1977
— 'Evolution, orchids and pandas', *New Scientist* (1978), *80*:700
— 'Perpetual youth', *New Scientist* (1979), *82*:832
— 'Human evolution', *New Scientist* (1979), *83*:738
GRANIT, R., *The Purposive Brain*, MIT Press, 1977
GRANT, V., *The Origin of Adaptations*, Columbia University Press, 1963
— *Modes of Speciation*, Columbia University Press, 1981 (2nd edn)
— *Organismic Evolution*, Freeman, New York, 1977
GRASSÉ, P. P., *Evolution of Living Organisms*, Academic Press, London and New York, 1977
GREGORY, M., SILVERS, A., and SUTCH, D. (eds.), *Sociobiology and Human Nature*, Jossey-Bass, San Francisco, 1978
GRUN, P., *Cytoplasmic Genetics and Evolution*, Columbia University Press, 1976
GURDON, J. B., 'Transplanted nuclei and cell differentiation', *Sc. Amer.* (1968), *219*(6)24
— 'How cells specialise', *Sc. Amer.* (1961), *205*(3)132
GUTFREUND, H. (ed.), *Biochemical Evolution*, Cambridge University Press, 1981
HADORN, E., 'Transdetermination in cells', *Sc. Amer.* (1968), *219*(5)110
HADZI, J., *The Evolution of the Metazoa*, Pergamon Press, Oxford, 1963
HAILMAN, J., 'How an instinct is learned', *Sc. Amer.* (1969), *221*(6)98
HALL, D. O., *et al.*, 'Role for ferredoxins in the origin of life and biological evolution', *Nature* (1971), *233*:136
HALLAM, A., 'The end of the Cretaceous', *Nature* (1979), *281*:430
HAMILTON, W. D., 'Extraordinary sex ratios', *Science* (1967), *156*:477
HANAWALT, P. C., and HAYNES, R. H., 'The repair of DNA', *Sc. Amer.* (1967), *216*(2)36
HANSON, E., 'Evolution of the cell from primordial living systems', *Qtly Rev. Biol.* (1966), *41*:1–12
HARBORNE, J. B. (ed.), *Biochemical Aspects of Plant and Animal Co-evolution*, Academic Press, London and New York, 1978
HARDY, Sir A., *The Living Stream*, Collins, London, 1965
HARGRAVES, R. B., 'Precambrian geologic history', *Science* (1976), *193*:363
HARPER, C. W., Jr, 'Origin of species in geologic time: alternatives to the Eldredge-Gould model', *Science* (1975), *190*:47

HARRIS, H., 'Enzyme polymorphism in man,' *Proc. Roy. Soc.*, London (1966), *164*:298

HARTMAN, P. E., and SUSKIND, S. R., *Gene Action*, Prentice Hall, Englewood Cliffs, NJ, 1965

HECHT, M., 'The role of natural selection and evolution rates on the origin of higher levels of organisation', *Syst. Zoo.* (1965), *14*:301

HEINRICH, B., and BARTHOLOMEW, G. A., 'Ecology of the African dung beetle', *Sc. Amer.* (1979), *241*(5)118

HENDERSON-SELLERS, A., and SCHWARTZ, A. W., 'Chemical evolution and ammonia in the early Earth's atmosphere', *Nature* (1980), *287*:526

HENDRICKS, S. R., 'How light interacts with living matter', *Sc. Amer.* (1968), *219*(3)174

HOFMANN, H. J., 'Canada's Precambrian fossils', *Geos* (Winter 1981)

HOUSE, M. R. (ed.), *The Origin of Major Invertebrate Groups*, Academic Press, London and New York, 1979

HOYLE, F., and WICKRAMASINGHE, C., *The Origin of Life*, University College of Cardiff Press, 1980

HSU, K. J., 'Terrestrial catastrophe caused by cometary impact at the end of the Cretaceous', *Nature* (1980), *285*:201

HURLEY, P. M., 'The confirmation of continental drift', *Sc. Amer.* (1968), *218*(4)52

HYAMS, J., 'Polarity of spindle microtubules', *Nature* (1980), *284*:402

JACOB, F., 'Evolution and tinkering', *Science* (1977), *196*:1161

JAMESON, D. L. (ed.), *Genetics of Speciation*, Bowden, Hutchinson and Ross, Stroudsberg, Pa, 1977

JANTSCH, E., *The Self-Organising Universe*, Pergamon Press, Oxford, 1980

JEPSEN, G. L., MAYR, E., and SIMPSON, G. G. (eds,.), *Genetics, Paleontology and Evolution*, Princeton University Press, 1949

JOHANSEN, K., 'Air-breathing fishes', *Sc. Amer.* (1968), *219*(4)102

JUKES, T. H., *Molecules and Evolution*, Columbia University Press 1966

— 'Silent nucleotide substitution and the molecular evolutionary clock', *Science* (1980), *210*:973

— and KING, J. L., 'Non-Darwinian evolution', *Science* (1969), *164*:788

— 'Evolutionary nucleotide replacements in DNA', *Nature* (1979), *281*:605

KAMEN, M., 'An universal molecule of living matter', *Sc. Amer.* (1958), *199*(2)77

KELLY, P., *Evolution and its Implications*, Mills & Boon, London, 1962

KERKUT, G. A., *Implications of Evolution*, Pergamon Press, Oxford, 1960 and 1973

KERR, R. A., 'Origin of life: new ingredients suggested', *Science* (1980), *210*:42

KIMBALL, A. P., and ORO, J. (eds.), *Prebiotic and Biochemical Evolution*, Elsevier, New York, 1971

KIMURA, M., 'The neutral theory of molecular evolution', *Sc. Amer.* (1979), *241*(5)94

— and OHTA, T., 'Protein polymorphism as a phase of molecular evolution', *Nature* (1971), *229*:467

KING, J. L., and JUKES, T. H., see Jukes and King

KING, M. C., and WILSON, A. L., 'Evolution at two levels in humans and chimpanzees', *Science* (1975), *188*:107
KNOLL, A. H., and BARGHOORN, E. S., 'Precambrian eukaryotic organisms', *Science* (1975), *190*:52
KOESTLER, A., *The Case of the Midwife Toad*, Hutchinson, London, 1971
— *Janus*, Hutchinson, London 1978
— and SMYTHIES, J. (eds.), *Beyond Reductionism*, Hutchinson, London, 1969
KOLATA, G. B., 'Repeated DNA sequences', *Science* (1973), *182*:1009
— 'Evolution of DNA', *Science* (1975), *189*:446
KORNBERG, R. D., 'Chromatin structure: a repeating unit of histones and DNA', *Science* (1974), *184*:868
— and THOMAS, J. O., 'Chromatin structure: oligomers of the histones', *Science* (1974), *184*:865
KURTEN, B., 'Continental drift and evolution', *Sc. Amer.* (1969), *220*(3)54
LACK, D., *The Course of Evolution Illustrated by Waterfowl*, Blackwell, Oxford, 1974
— 'Darwin's finches', *Sc. Amer.* (1953), *188*(4)66
LAND, M., 'Animal eyes', *New Scientist* (1979), *84*:10
LASKEY, R. A., and EARNSHAW, W. C., 'Nucleosome assembly', *Nature* (1980), *286*:763
LEEPER, G. W. (ed.), *The Evolution of Living Organisms*, University of Melbourne Press, 1962
LEHMAN, J. P., *The Proofs of Evolution*, Gordon and Cremonesi, New York, 1977
LEVIN, D. A., 'The nature of plant species', *Science* (1979), *204*:381
LEVINE, R. P., 'The mechanisms of photosynthesis', *Sc. Amer.* (1969), *221*(6)58
LEVINS, R., *Evolution in Changing Environments*, Princeton University Press 1968
LEVI-Setti, R., *Trilobites: a photographic atlas*, University of Chicago Press, 1975
LEWIN, R., 'Genes in action', *New Scientist* (1976), *70*:623
— 'Overlapping genes', *New Scientist* (1976), *72*:148
LEWIS, H., 'Catastrophic selection as a factor in speciation', *Evolution* (1962), *16*:257
— 'Speciation in flowering plants', *Science* (1966), *152*:167
— and RAVEN, P., 'Rapid evolution in Clarkia', *Evolution* (1958), *12*:319
LEWIS, J. (ed.), *Beyond Chance and Necessity*, Garnstone Press, London, 1974
LEWONTIN, R., 'Adaptation', *Sc. Amer.* (1978), *239*(3)156
— 'Analysis of variance and analysis of cause', *Amer. J. Human Genetics* (1974), *26*:400–411
— *The Genetic Basis of Evolutionary Change*, Columbia University Press, 1974
LIMBAUGH, C., 'Cleaning symbiosis', *Sc. Amer.* (1961), *205*(2)42
LOCKE, M. (ed.), *Major Problems in Developmental Biology*, Academic Press, London and New York, 1965
— *The Emergence of Order in Developing Systems*, Academic Press, London and New York, 1968
LOOMIS, W. F., 'The sex gas of Hydra', *Sc. Amer.* (1959), *200*(4)145

LORENZ, K., *Evolution and Modification of Behaviour*, University of Chicago Press, 1967

LOVELOCK, J. E., *Gaia: A New Look at Life on Earth*, Oxford University Press, 1979

LOWE, D., 'Stromatolites 3400 Myr-old from the Archean of Western Australia', *Nature* (1980), *284*:441

LUCY, J. A., and GLAVERT, A., 'Structure and assembly of macromolecular lipid complexes composed of globular micelles', *J. Mol. Biol.* (1964), *8*:727

MACBETH, N., *Darwin Retried*, Garnstone Press, London, 1974

— *The Differentiation of Cells*, Arnold, London, 1977

MADEN, M., and TURNER, R. N., 'Supranumerary limbs in the axolotl', *Nature* (1978), *273*:232

MAIO, J. J., 'Predatory fungi', *Sc. Amer.* (1958), *199*(1)67

MANIATIS, T., and PTASHNE, M., 'A DNA operator-repressor system', *Sc. Amer.* (1976), *234*(1)64

MARGULIS, L., *Origin of Eukaryotic Cells*, Yale University Press, 1970

— 'Symbiosis and evolution', *Sc. Amer.* (1971), *225*(2)48

— (ed.) Origins of Life (conference on), Gordon and Breach, New York, London, Paris, 1970

— WALKER, J. C. G., and RAMBLER, M., 'Reassessment of roles of oxygen and ultraviolet light in Precambrian evolution', *Nature* (1976), *264*:620 –24

MARKERT, C. L., 'Mechanism of cellular differentiation' from J. A. Moore (ed.), *Ideas in Modern Biology*, p.231

— *et al.*, 'Evolution of a gene', *Science* (1975), *189*:102

MARTIN, R. D., 'Phylogenetic reconstruction versus classification: the case for clear demarcation', *Biologist* (1981), 28(3)

MARX, J., 'Gene structure; more surprising developments', *Science* (1978), *199*:517

MASON, B., 'Organic matter from space', *Sc. Amer.* (1963), *208*(3)43

MATTHEWS, R. W. and MATTHEWS, J. R., *Insect Behaviour*, Wiley-Interscience, Chichester, 1978

MAUGH, T. H., II, 'Phylogeny: are methanogens a third class of life?', *Science* (1977), *198*:812

MAURO, A., and STEN-KNUDSEN, O., 'Light-evoked impulses from extra-ocular photo-receptors in the squid Todarodes', *Nature* (1972), *237*:342

MAY, R. M., 'Group selection' (letter), *Nature* (1975), *254*:485

MAYNARD SMITH, J., *The Theory of Evolution*, Penguin, Harmondsworth, 1958

— '"Haldane's Dilemma" and the rate of evolution', *Nature* (1968), *219*:1114

— 'Genes and race', *Nature* (1981), *289*:742

— and HOLLIDAY, R. (eds.), 'The evolution of adaptation by natural selection', *The Royal Society*, 1980 (symposium)

MAYR, E., *Animal Species and Evolution*, Belknap Press, Harvard University Press, 1963 (1973 printing)

— 'The nature of the Darwinian revolution', *Science* (1972), *176*:981

— 'Darwin and natural selection', *Amer. Scientist* (1977), *65*:321
— *Systematics and the Origin of Species*, Columbia University Press, 1942; Dover, New York, 1964
— and PROVINE, William B., *Evolutionary Synthesis: perspectives on the unification of biology*, Harvard University Press, 1980
MCCAPRA, F., and HART, R., 'The origins of marine bioluminescence', *Nature* (1980), *286*:660
MCFARLAND, D. (ed.), *The Oxford Companion to Animal Behaviour*, Oxford University Press, 1981
MCGOWAN, D., and BAKER, A. J., 'Common ancestry for birds and crocodiles?', *Nature* (1981), *289*:97
MEDVEDEV, Z. A., *The Rise and Fall of T. D. Lysenko*, Columbia University Press, 1969
MEKLER, L. B., 'Mechanism of biological memory', *Nature* (1967), *215*:481
MILLER, K. R., 'The photosynthetic membrane', *Sc. Amer.* (1979), *241*(4)100
MILLER, S., and ORGEL, L. E., *The Origins of Life on Earth*, Prentice Hall, Englewood Cliffs, NJ, 1974
MILNE, L. J., and MILNE, M. J., 'Criminal Courtships', *Sc. Amer.* (1950), *183*(1)52
MILNER, A., 'Triassic extinction of Jurassic vacuum?', *Nature* (1977), *265*:402
— 'Early assemblage of Carboniferous amphibians', *Nature* (1977), *265*:495
— 'Flamingos, stilts and whales', *Nature* (1981), *289*:347
MILTON, D., 'Drifting organisms in the Pre-Cambrian sea', *Science* (1966), *153*:293
MODELL, W., 'Horns and antlers', *Sc. Amer.* (1969), *220*(4)114
MOORE, J. A. (ed.), *Ideas in Modern Biology*, Proc. Int. Congr. Zoo., 1965
MOORE, R. C. (ed.), *Treatise on Invertebrate Paleontology*, Geol. Soc. of Amer. & Univ. of Kansas, 1969
MOORHEAD, P. S., and KAPLAN, M. M. (eds.), *Mathematical Challenges to the Neo-Darwinian Interpretation of Evolution*, Wistar Inst., Philadelphia, 1967
MOSCONA, A. A., 'Tissues from dissociated cells', *Sc. Amer.* (1959), *200*(5)132
NEFF, W. D. (ed.), *Contributions to Sensory Physiology Vol. 2*, Academic Press, London and New York, 1967
NEWELL, N. D., 'Crises in the history of life', *Sc. Amer.* (1963), *208*(2)76
NIERLICH, D. P., RUTTER, W. J., and FOX, C. F. (eds.), *Molecular Mechanisms in the Control of Gene Expression*, Academic Press, London and New York, 1976
NISBET, E., 'Archaean stromatolites and the search for the earliest life', *Nature* (1980), *284*:395
— and PILLINGER, C., 'In the beginning', *Nature* (1981), *289*:11
NURSOLL, J. R., 'On the origin of major groups of animals', *Evolution* (1962), *16*:118
OGLE, R. B., *Animals and their Camouflage*, Foulsham, Slough, 1959
OHNO, S., 'Conservatives win the evolutionary race', *New Scientist* (1974), *61*:412

— *Evolution by Gene Duplication*, Springer Verlag, New York, 1970
— 'Simplicity of mammalian regulatory systems inferred by single gene determination of sex phenotypes', *Nature* (1971), *234*:134
OLSON, E., 'The evolution of mammalian characteristics', *Evolution* (1959), *13*:344
OLSON, J. M., 'The evolution of photosynthesis', *Science* (1970), *168*:438
ORGEL, L. E., *The Origins of Life: Molecules and Natural Selection*, Wiley-Interscience, Chichester, 1973
OWEN *et al.*, 'Enhanced CO2 greenhouse to compensate for reduced solar luminosity on early earth', *Nature* (1979), *277*:640
PATTEE, H., 'On the origin of macromolecular sequences', *Biophys. J.* (1960–61), *1*:683
PAUL, J., 'A general theory of chromosome structure and gene activation in eukaryotes', *Nature* (1972), *238*:444
PEARSON, R., *Climate and Evolution*, Academic Press, London and New York, 1978
PFLUG, H. D., and JAESCHYKE-BOYER, H., 'Combined structural and chemical analysis of 3800 Myr-old microfossils', *Nature* (1979), *280*:483
PIAGET, J., *Biology and Knowledge*, Edinburgh University Press, 1971
— *Behaviour and Evolution*, Routledge and Kegan Paul, London, 1979
POLEZHAEV, L. V., *Organ Regeneration in Animals*, Thomas, Springfield, Ill. 1972
PRIGOGINE, I., *From Being to Becoming*, Freeman, San Francisco, 1980
— 'Irreversibility as a symmetry-breaking process', *Nature* (1973), *246*:67
PTASHNE, M., and GILBERT, W., 'Genetic repressors', *Sc. Amer.* (1970), *222*(6)36
PURCELL, J. E., 'Influence of siphonophore behaviour upon their natural diets: evidence for aggressive mimicry', *Science* (1980), *209*:1045
RAUP, D. M., 'Taxonomic diversity during t. Phanerozoic', *Science* (1972), *177*:1065
RAVEN, P. H., 'A multiple origin for plastids and mitochondria', *Science* (1970), *169*:641–6
REGAL, P., 'Ecology and evolution of flowering plant dominance', *Science* (1977), *196*:622
REID, G. C., *et al.*, 'Effects of intense stratospheric ionisation events', *Nature* (1978), *275*:489
REINERT, J., and URSPRUNG, H., *Origin and Continuity of Cell Organelles*, Springer Verlag, Berlin, 1971
RENDEL, J. M., 'Control of developmental processes' in DRAKE, J. W. (ed.), p. 341
RENSCH, B., *Evolution above the Species Level*, Methuen, London, 1959
RIDLEY, M., 'Evolution and gaps in the fossil record', *Nature* (1980), *286*:444
RIEDL, R., *Order in Living Organisms*, Wiley-Interscience, Chichester, 1978
RINGO, J. M., 'Why 300 species of Hawaiian drosophila?', *Evolution* (1977), *31*:694
ROBERTSON, M., 'Beads of life', *New Scientist* (1979), *83*:8
ROE, A., and SIMPSON, G. G. (eds.), *Behavior and Evolution*, Yale University Press, 1958

ROGERS, J., 'Evolution of eukaryotes', *New Scientist* (1976), *71*:333
ROMER, A. S., 'Major steps in vertebrate evolution', *Science* (1967), *158*:1629
RODYN, D. B., and WILKIE, D., *The Biogenesis of Mitochondria*, Methuen, London, 1968
RUSSELL, E. S., *The Directiveness of Organic Activities*, Cambridge University Press, 1945
SAGER, R., *Cytoplasmic Genes and Organelles*, Academic Press, London and New York, 1972
SANGER, E., 'Amended dogma', *Sc. Amer.* (1977), *236*(5)50
SARGENT, T. D., 'Background selection of the pale and melanic forms of the cryptic moth, Philagalia Titea', *Nature* (1969), *222*:585
SATIR, P., 'How cilia move', *Sc. Amer.* (1974), *231*(4)44
SAUNDERS, P. T., and BAZIN, M. J., 'Stability of complex ecosystems', *Nature* (1975), *256*:120
SCHIMKE, R. T., 'Gene amplification and drug resistance', *Sc. Amer.* (1980), *243*(5)50
SCHMALHAUSEN, I. I., *The Origin of Terrestrial Vertebrates*, Academic Press, London and New York, 1968
SCHOFFENIELS, E. (ed.), *Molecular Evolution (vol. 2): Biochemical Evolution and the Origin of Life*, N. Holland, New York, 1971
SCHOPF, J. W., 'On the development of metaphytes and Metazoans', *J. Paleontol.* (1973), *47*:1
— 'The evolution of the earliest cells', *Sc. Amer.* (1978)), *239*(3)84
— and OEHLER, D. Z., 'How old are the eukaryotes?', *Science* (1976), *193*:47
SCHOPF, T. J. M., *Models in Paleobiology*, Freeman, San Francisco, 1972
SCOTT, J., 'Natural selection in the primordial soup', *New Scientist* (1981), *89*:153
SEELY, M. K., and HAMILTON III, W. J., 'Fog catchment sand trenches constructed by tenebrionid beetles from the Namib desert', *Science* (1976), *193*:484
SENIOR, E., *et al.*, 'Enzyme evolution in a microbial community growing on the herbicide Dalapon', *Nature* (1976), *263*:476
SEUANEZ, H. N., *The Phylogeny of Chromosomes*, Springer Verlag, Berlin and New York, 1979
SHELDRAKE, R., *A New Science of Life*, Blond and Briggs, London, 1981
SHUSTER, G. W., and THORSON, R. (eds.), *Evolution in Perspective*, University of Notre Dame Press, 1970
SIMPSON, G. G., *Splendid Isolation: the curious history of South American mammals*, Yale University Press, 1980
— *Tempo and Mode in Evolution*, Harvard University Press, 1944
SINNOTT, E. W., *The Problem of Organic Form*, Yale University Press, 1963
SMIT, J., and HERTOGEN, J., 'An extraterrestrial event at the Cretaceous-Tertiary boundary', *Nature* (1980), *285*:198
SMITH, H., *From Fish to Philosopher*, Little Brown, Boston, 1953
SMITH, J. E., *et al.*, *The Invertebrate Panorama*, Weidenfeld and Nicolson, London, 1971
SOKAL, R. R., and CONELLO, T. J., 'The biological species concept: a critical evaluation', *Amer. Nat.* (1970), *104*:127

STACK, S. M., and BROWN, W. V., 'Somatic pairing, reduction and recombination: an evolutionary hypothesis of meiosis', *Nature* (1969), *222*:1275

STAHL, B., *Vertebrate History: problems in evolution*, McGraw-Hill, Maidenhead, 1974

STANLEY, S. M., *Macro-evolution: pattern and process*, Freeman, San Francisco, 1979

— 'An evolutionary theory for the sudden origin of multicellular life in the late Pre-Cambrian', *Proc. Nat. Acad. of Sciences, USA*, (1973), *70*:486

— 'Clades versus clones in evolution: why we have sex', *Science* (1975), *190*:382

— and HARPER, C. W., Jr, 'Stability of species in geologic time', *Science* (1976), *192*:267

STEBBINS, G. L., *The Basis of Progressive Evolution*, University of North Carolina Press, 1969

STEELE, E. (ed.), *Genetic Selection and Adaptive Evolution*, Williams and Wallace, Toronto, 1979

STEIN, G. S., *et al.*, 'Non-histone chromosal proteins and gene regulation', *Science* (1974), *183*:817

— 'Chromosal proteins and gene regulation', *Sc. Amer.* (1975), *232*(2)46

STUERMER, W., 'Soft parts of cephalopods and trilobites: some surprising results of x-ray examination of Devonian slates', *Science* (1970), *170*:1300

TAX, S. (ed.), *Evolution After Darwin*, Chicago University Press, 1960

TAYLOR, C., *The Explanation of Behaviour*, Routledge, London, 1964

TAYLOR, T. G., 'How an eggshell is made', *Sc. Amer.* (1970), *222*(3)89

TEMIN, M. M., 'The DNA provines hypothesis', *Science* (1976), *192*:1075

TEMPLE, S. A., 'Plant animal mutualism', *Science* (1977), *197*:887

TEMPLETON, A., 'Why 300 species of Hawaiian drosophila?', *Evolution* (1979), *33*:513

THODAY, J. M., 'Non-Darwinian "evolution" and biological progress', *Nature* (1975), *255*:675

THOMPSON, J. N., Jr, and WOODRUFF, R. C., 'Mutator genes: pacemakers of evolution', *Nature* (1978), *274*:317

THOMPSON, K. S., 'Explanation of large scale extinction of lower vertebrates', *Nature* (1976), *261*:578

THORNTON, C. S., and BROMLEY, S. C., *Vertebrate Regeneration*, Dowden, Hutchinson and Ross, Stroudsberg, Pa, 1974

THORPE, W. H., *Purpose in a World of Chance*, Oxford University Press, 1978

— 'Duet-singing birds', *Sc. Amer.* (1973), *229*(2)70

THULBORN, R. A., 'Origins and evolution of ornithischian dinosaurs', *Nature* (1971), *234*:75

TOATES, F. M., *Animal Behavior – a systems approach*, Wiley, Chichester, 1980

TOMKINS, G. M., *et al.*, 'Control of specific gene expression in higher organisms', *Science* (1969), *166*:1474

TORREY, T. W., *Morphogenesis of the Vertebrates*, Wiley, Chichester, 1971

TOWE, K., 'Oxygen collagen priority of the early Metazoans', *Proc. Nat. Acad. Sc. USA* (1970), *65*:781

— 'Early Pre-Cambrian oxygen: a case against photosynthesis', *Nature* (1978), *274*:657

— 'Trilobite eyes: calcified lenses in vivo', *Science* (1973), *179*:1007

VALENTINE, J. W., 'The evolution of multicellular plants and animals', *Sc. Amer.* (1978), *239*(3)104

VAN ABEELEN, J. H. F. (ed.), *The Genetics of Behaviour*, N. Holland/Elsevier, New York, 1974

VAN BERGEIJK, W., 'The evolution of vertebrate hearing' in NEFF, W. D. (ed.), pp. 1–50

VAN VALEN, L., 'The history and stability of atmospheric oxygen', *Science* (1971), 171:439

WADDINGTON, C. H., *Organisers and Genes*, Cambridge University Press, 1947

— *Evolution of an Evolutionist*, Edinburgh University Press, 1975

— *The Nature of Life*, Allen and Unwin, London, 1961

— 'The theory of evolution today' contr. to KOESTLER and SMYTHIES (eds.), *Beyond Reductionism*, pp. 357–74. Discussion pp. 375–95. Hutchinson, London, 1969

WAHLERT, G., 'The role of ecological factors in the origin of higher levels of organisation', *Syst. Zoo.* (1965), *14*:288

WALD, G., 'The significance of vertebrate metamorphosis', *Science* (1958), *128*:1481

— 'Life and light', *Sc. Amer.* (1959), *201*(4)92

WALLACE, R., *The Ecology and Evolution of Animal Behavior*, Goodyear, Pacific Palisades, 1973

WALLS, G. L., *The Vertebrate Eye and Its Adaptive Radiation*, Cranbrook Inst. of Science, Bloomfield Hills, Mich., 1942

— 'The visual cells and their history', *Biol. Symp.* (1962), *7*:203

WALTER, M. R., BUICK, R., and DUNLOP, J., 'Stromatolites 3400–3500 Myr-old from the North Pole area, Western Australia', *Nature* (1980), *284*:443

WARDLAW, C. W., *Organisation and Evolution in Plants*, Longmans, London, 1965

WARREN, A., 'Jurassic labyrinthodont', *Nature* (1977), *265*:436

WATSON, D. M. S., 'The origin of frogs', *Trans. Roy. Soc. Edin.* (1941), *60*(i)195

— *Paleontology and Modern Biology*. Silliman Lectures, 1937. Yale University Press, 1951

WEISS, P. A. (ed.), *Hierarchically Organised Systems in Theory and Practice*, Hafner Publishing, New York, 1971

WESTOLL, T. S. (ed.), *Studies on Fossil Vertebrates* (presented to D. M. S. Watson), University of London Press, 1958

WHEATLEY, D., 'How cells select amino-acids', *New Scientist* (1980), 87:294

WHETSTONE, K., and MARTIN, L., 'New look at the origin of birds and crocodiles', *Nature* (1979), *279*:234

WHITAKER, T. H., 'New concepts of kingdoms of organisms', *Science* (1969), *163*:150

WHITE, M. J. D., *Modes of Speciation*, Freeman, San Francisco, 1978
WHYTE, L. L., *Internal Factors in Evolution*, Tavistock Publications, London, 1965
WICKLER, P., *Mimicry in Plants and Animals*, Weidenfeld and Nicolson, London, 1968
WIEDEMANN, J., 'Heteromorphs and ammonite extinction', *Biol. Revs.* (1969), *44*:563
WILKIE, D., *The Cytoplasm in Heredity*, Methuen, London, 1964
WILKINSON, J. F., 'Methanogenic bacteria: a new primary kingdom?', *Nature* (1978), *271*:707
WILLIAMS, G. C., *Adaptation and Natural Selection*, Princeton University Press, 1966
—*Sex and Evolution*, Princeton University Press, 1975
WILLIAMSON, P. G., 'Paleontological documentation of speciation in Cenozoic molluscs from Turkana Basin', *Nature* (1981), *293*:437
WILLIAMSON, R., 'DNA insertions into gene structure', *Nature* (1977), *270*:295
WILLIS, J. C., *The Course of Evolution by Divergence of Mutation*, Cambridge University Press, 1940
— *Age and Area*, Cambridge University Press, 1922
WILLIS, R. A., *The Borderland between Embryology and Pathology*, Butterworth, London, 1958
WILLMER, E. N., *Cytology and Evolution (2nd ed.)*, Academic Press, New York and London, 1970
WILLIS, C., 'Genetic load', *Sc. Amer.* (1970), *222*(3)98
WILSON, A. C., *et al.*, 'The importance of gene re-arrangement', *Proc. Nat. Acad. Sc. USA* (1974), *71*:3028
— 'Social structuring of mammal populations and rate of chromosal evolution', *Proc. Nat. Acad. Sc. USA* (1975), *72*:5061
— 'Biochemical evolution', *Ann. Rev. Biochem.* (1977), *46*:573
WILSON, E. O., *On Human Nature*, Harvard University Press, 1978
— 'The nature of human nature', *New Scientist* (1978), *80*:20
WINGERSON, L., 'Pattern formation in nature', *New Scientist* (1981), *90*:566
WITT, P. N., 'Do we live in the best of all possible worlds?', *Pers. in Biol. and Med.* (1965), *8*:475
WOESE, C., and FOX, G. E., 'The concept of cellular evolution', *J. Molec. Evol.* (1977), *10*:1–6
— 'Phylogenetic structure of the prokaryotic domain: the primary kingdoms', *Proc. Nat. Acad. Sc. USA* (1977), *74*:5088
WOESE, C. R., and GUPTA, R., 'Are archaebacteria merely derived prokaryotes?', *Nature* (1981), *289*:95
WOLFF, E., *The Relationship between Experimental Embryology and Molecular Biology*, Gordon and Breach, New York, London, Paris, 1967
WOLSKY, M. and A., *The Mechanism of Evolution: a new look at old ideas*, Karger, Basel, 1976
WOOD JONES, F., *Trends of Life*, Arnold, London, 1953
WOODRUFF, H. B., 'Natural products from microorganisms', *Science* (1980), *208*:1225

YONGE, C. M., 'Giant clams', *Sc. Amer.* (1975), *232*(4)96
ZELENY, M. (ed.), *Autopoiesis, Dissipative Structures and Spontaneous Social Orders*, Westview Press, Boulder, Col., 1980

Sources

Charles W. Andrews: *A Descriptive Catalogue of the Marine Reptiles of the Oxford Clay*, 1910: 66 top
Ardea, London: 73, 96, 111, 113, 120, 123, 124, 149, 153
David L. Clark: *Fossils, Paleontology and Evolution*, William C. Brown Co., Dubuque Iowa, 1976: 26, 125 top
Edward Clodd: *The Story of Creation*, 1888: 7
Edwin H. Colbert: *Evolution of the Vertebrates*, John Wiley & Sons, 1969, redrawn from Gerhard Heilmann: *The Origin of Birds*, 1927: 68
Sir Gavin de Beer: *The Atlas of Evolution*, Thomas Nelson Ltd. and B.V. Uitgeversmaatschappij Elsevier, 1964: 30, 31, 32, 100, 159, 205
George F. Eaton: *Osteology of Pterandon*, 1910: 121 bottom
Mary Evans Picture Library: 16, 28
William K. Gregory: *Evolution Emerging*, Macmillan, New York, 1951: 62, 118
By permission of Erik Jarvik, from *The Scientific Monthly*, Vol. 80, 1955: 60
Bjorn Kurten: *The Age of the Dinosaurs*, Weidenfeld & Nicolson, 1968: 58, 59, 121 top
J. P. Lehmann: *The Proofs of Evolution*, 1977 (originally published by Presses Universitaires de France): 76
W. D. Matthew and W. Granger: *Bulletin of the American Museum of Natural History* No. 37, 1917: 72
W. S. McKerrow (ed.): *The Ecology of Fossils*, Duckworth, 1978 (drawing by E. Winson): 142
Nature: frontispiece
J. R. Norman: *A History of Fishes* (3rd edition by P. H. Greenwood), Ernest Benn, 1975: 25, 66 bottom, 110
Henry Fairfield Osborn: *The Titanotheres of Ancient Wyoming*, 1929: 126
M. F. Perutz: "The Hemoglobin Molecule", *Scientific American* November 1964. Copyright © 1964 by Scientific American, Inc. All rights reserved: 107
George J. Romanes: *Darwin and After Darwin*, 1892: 67, 69, 128
Alfred S. Romer: *The Procession of Life*, Weidenfeld & Nicolson, 1968: 61, 63, 117, 178, 187
William B. Scott: *A History of Land Mammals in the Western Hemisphere*, 1913: 125 bottom
George Gaylord Simpson: *Life of the Past*, Yale University Press, 1953: 29
Samuel Wendell Williston: *Osteology of the Reptiles*, 1925: 65

Index

Items in italic indicate illustrations.

About the Author

Gordon Rattray Taylor died in December 1981, soon after he had completed *The Great Evolution Mystery.* He was educated at Radley College and Trinity College, Cambridge. Until World War II he worked on the *Morning Post* and the *Daily Express.* He then worked in the BBC Monitoring Service, on European news and as intelligence officer in the Psychological Warfare Division of SHAEF. Subsequently he was Chief Science Advisor, BBC Television, and editor of the *Horizon* series; he worked on the *Eye on Research* series, which won the World's Best Science TV Programme Award, and *Machines Like Men,* which won the Brussels Award.

He was the author of fifteen books including *The Biological Time Bomb; The Doomsday Book; Rethink: A Paraprimitive Solution; How to Avoid the Future;* and *The Natural History of the Mind.*

Mr. Taylor was married and had two daughters.